BUSINESS TELECOMMUNICATIONS

BUSINESS TELECOMMUNICATIONS

Data Communications in the Information Age

FRANK GREENWOOD
Southeastern Massachusetts University

MARY M. GREENWOOD
Center for Productivity, Inc.

ROBERT E. HARDING
Polytop Corporation

Wm C. Brown Publishers
Dubuque, Iowa

Company Trademark

AT&T is a registered trademark. dBase III Plus is a registered trademark of Ashton-Tate Inc. Digital and Ethernet are registered trademarks of Digital Equipment Corporation. Federal Mogul is a registered trademark of Federal Mogul Corporation. GM is a registered trademark of General Motors Corporation. GE is a registered trademark of General Electric Company. The following are registered trademarks of International Business Machine Corporation: IBM 360/50; IBM 650; IBM 1001; IBM 1130; IBM 3033; IBM 3084 Model 0X; IBM Personal Computer; IBM Personal Computer XT; IBM Personal System/2. Infotron is a registered trademark. Lotus is a registered trademark of Lotus Development Corporation. MCI is a registered trademark. Microsoft is a registered trademark. NYNEX is a registered trademark. Olivetti is a registered trademark. Paradyne is a registered trademark. Rolm is a registered trademark. Telenet is a registered trademark. Teletype is a registered trademark. TopView is a registered trademark. Tymshare is a registered trademark. VisiCalc is a registered trademark. Xerox is a registered trademark.

To Our Parents

CONTENTS

CHAPTER 3
DIGITAL COMMUNICATIONS   47

CHAPTER 4
SOME BUSINESS TELECOMMUNICATIONS COMPONENTS   61

CHAPTER 5
SOFTWARE   79

PREFACE

Telecommunications and computers are the crucial technologies of the information age and, as you'll see in this book, they are both part of the same information-processing continuum. IBM's and AT&T's efforts to penetrate both the computer and telecommunications markets provide evidence of this.

Since the U.S. Census Bureau bought the first commercial computer in 1951, we have created a vast computer-related technology. It was apparent early on that separate computer applications (such as accounting, data base, and process control programs) are often more productive when connected and allowed to share information, hardware, and software. So the need for business telecommunications has grown along with the number of computer applications.

Many organizations now have isolated "islands" of automation—separate (and often incompatible) applications for word processing, general ledger, and inventory record keeping. The technical challenge of telecommunications is to connect these islands. Linking engineering, production, and management into one computerized information system, for example, can improve product quality, productivity, and decision making. And this makes us more competitive.

Telecommunications makes office and factory automation a reality. The data, document, or message that we cannot readily communicate is diminished in value. Using computers and telecommunications to collect, analyze, and distribute information can provide the edge in everything from product development to marketing, and the technologies and skills now exist to make information a competitive weapon in the marketplace. Computers are no longer just for accountants.

In the broadest sense, telecommunications includes all of the hardware and software necessary for the transmission and reception of information. By *information*, we mean voice, printed messages, graphic displays, images, tone signals, digital and analog codes, and any other impulse variations that are transmitted from one location to another. Information travels over such media as wires, cables, fiber optics, radio waves, microwaves, and satellites.

This book, however, uses a narrower definition. Business telecommunications is here concerned with the electronic transmission of computer data. Consequently, we are only incidentally interested in image transmission via cable television, for example.

Chapter 1 begins by explaining the importance of telecommunications, and Chapter 2 then takes a general look at how a terminal talks to a computer over the public telephone system. Chapter 3 discusses how we are moving toward digital telecommunications as we replace and expand our existing telephone system.

Some business telecommunications concepts are discussed in Chapter 4, including public and private networks. Chapter 5 considers software concepts, such as line control, that are useful in telecommunications.

Chapter 6 shows how a logical communication channel is established and how transmission occurs. Having reviewed software and protocols, we can tackle network architecture in Chapter 7. We then turn to local area networks in Chapter 8, and standards, regulation, and divestiture are covered in Chapter 9.

Chapter 10 provides a perspective on why telecommunications matters to executives. Network design is considered in Chapter 11, and errors and security are the main topics in Chapter 12.

We hope this book will be an easy introduction to telecommunications for business people and students without any background in the subject. To accustom readers to the literature, articles on telecommunications from magazines and journals are presented at the end of each chapter. There are also two case studies that deal with the impact of telecommunications on the business environment.

To simplify the reader's task, each chapter begins with a chapter outline and a brief list of objectives and ends with a summary and some review questions. There is a glossary at the end of the book, defining many of the terms used.

To assist the instructor, there is a separate instructor's manual which includes teaching suggestions, chapter outlines, answers to the review questions, as well as true/false and multiple-choice questions and answers.

In a four-year undergraduate curriculum, this book can be the basis for an introductory data communications course. It also meets computer science requirements established by the Association for Computing Machinery and could serve as the basis for teaching IS6 — Data Communications, Networks, and Distributed Processing. This text can also be used for data communications courses in a two-year data-processing program.

Acknowledgments

Our intent has been to create a competitive, quality product. We therefore acknowledge the critical contributions made by a number of individuals who reviewed our work at various stages: Dale Buchholz, DePaul University; Cynthia E. Johnson, Bryant College; Kenneth P. Johnson, Grand Valley State College; Rose Laird, Northern Virginia Community College; Gerarld R. Lamphere, Detroit College of Business; and Curtis Rawson, Kirkwood Community College.

BUSINESS TELECOMMUNICATIONS

1 INTRODUCTION TO TELECOMMUNICATIONS

BACKGROUND   **THIS BOOK**
INTEGRATION   **THE CASE STUDIES**
BUSINESS
    TELECOMMUNICATIONS

AFTER READING THIS CHAPTER, YOU WILL UNDERSTAND THAT:

- Telecommunications is now the premier form of information technology.
- Information technology has become cheaper and more powerful, providing affordable solutions to our productivity problems.
- The International Standards Organization has recommended a model, or theoretical framework, which is necessary for two machines to communicate. This model tends to define parameters for communication hardware and software.
- In addition, the Brisfield Company case study is presented. This case study will be referred to in most chapters.

Telecommunications is now the premier form of information technology: the data, document, or message that cannot readily be communicated diminishes in value. IBM's part-ownership of MCI, a long-distance telephone company, and AT&T's interest in Olivetti, a business machines manufacturer, give evidence that computers and communications are parts of the same continuum.

When an office has several personal computers, it is usually productive to connect them so they can exchange data and share the same information. When a manufacturing plant has a number of control machines, robots, and the like, they are often connected into a single network.

Such **local area networks (LANs)** are becoming common. The emerging factory standard is the Manufacturer's Automation Protocol (MAP), which is embraced by GM and others. A major office LAN is Ethernet, the standard used by Digital Equipment Corporation (DEC) and Xerox, for example.

Long-distance communication has used satellites since the 1960s. Orbiting some 22,500 miles above the equator as the earth spins on its axis, they remain "fixed" over one spot on the earth's surface. Surface distances are of little importance to satellites: communicating between Los Angeles and San Francisco via satellite is no different than between Los Angeles and Tokyo, and costs the same.

Another communications development is the Integrated Services Digital Network (ISDN) being implemented by NYNEX and other telephone companies, which sends all messages by digital codes. But instead of the dots and dashes of international Morse code, digital signals are composed of **binary** digits: *zeros* and *ones*. With the Integrated Services Digital Network, everything goes digitally: voice, data, and video. Chips and other electronic developments make digital circuits cheaper, more accurate, and easier to maintain than the old analog circuits once used in telephone communications.

Fiber optics are also becoming important in communications. Instead of using electricity to carry the code—the binary digits—fiber optics use light as the carrier, with lasers frequently used as the light source. Fiber optic circuits have a huge capacity to carry information, which makes light a promising technology in computers and communications. Some labs are experimenting with computers that use light internally instead of electricity.

We begin to see that computers and communications are both essential parts of automation. As we learn how to connect them, we will cut costs, reduce clerical effort and paperwork, raise quality, and improve productivity. Local area networks, satellites, Integrated Services Digital Networks, and fiber optics are examples of how information is communicated in ways that help us to be more productive.

This chapter presents some background, preparing the reader for the material to come on the technologies themselves.

——— BACKGROUND ———

During the 1950s, when computers first began to be used commercially, they were big and expensive. Their slow electronic circuits used vacuum tubes. By the early 1960s, transistors had replaced vacuum tubes inside computers, and by 1965, integrated circuits were being used. These gave way to *large-scale integrated (LSI)* circuits (quarter-inch square chips of silicon containing a large number of integrated circuits) in the 1970s and *very large scale integration (VLSI)* soon followed: chips with thousands of transistors, capable of storing about one million binary digits.

Now, a *VLSI random access memory* chip the size of your toenail can hold 10,000 telephone numbers. (Random access memory, or RAM, is memory that provides temporary storage for data and program instructions.) And a microprocessor smaller than your finger can outperform the first generation of mainframes, which filled big rooms (see Figure 1–1).

The technology has thus become cheaper, more powerful, and much smaller. This means that for the first time we have affordable technical solu-

FIGURE 1–1

This Intel micro-computer is tiny compared to the finger upon which it rests.

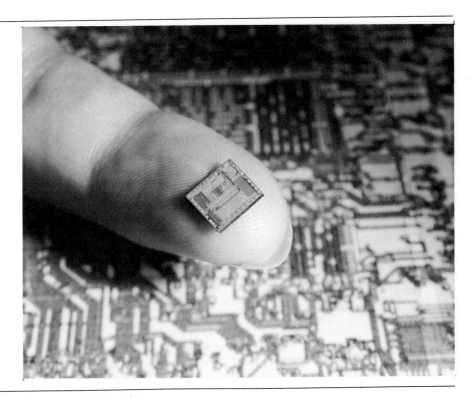

Courtesy Intel Corporation.

tions to our productivity problems. But these solutions have to be integrated to be effective, and it is the telecommunications network that achieves this integration.

Early computer centers tended to be isolated from their organizations by their unique technologies. Because of personnel and equipment costs, these computer centers usually did all the electronic data processing an organization needed; that is, costs were so high that all the organization's work was channeled through one center to maximize the processing volume and to reduce the cost per unit processed. Punch cards were used, and they were processed in batches, so whatever work you wanted run was done via these cards.

Time-sharing evolved, so clerks or managers could do a limited amount of their own work at a terminal without relying on the computer center technicians. For example, company data files stored in a central computer at the other end of the state could be accessed via a local terminal that shared time on the computer with other users. From this, **distributed data processing (DDP)** was developed whereby smaller computers could communicate with one or more central computers in a coordinated fashion. This usually provides some local independence—for instance, the separate processors can run application programs with their own data.

While **mainframes** (large computers with high data storage capacity and capable of processing data very fast) matured within the data-processing department, other departments were becoming more sensitive to the new technologies. Word processing was an early example. Word processing hardware was relatively expensive at first, which often made word processing centers advisable. Document creation could thus be centralized, keeping the hardware and the operators busy. As technological developments shrank the circuitry and allowed many more functions on smaller silicon chips, the equipment dropped in price and became more powerful, so there was less need for it to work all the time and less reason for centralized word processing. Many secretaries now have their own word processors, which can be idle much of the time without anyone worrying about cost.

Computer centers are busy places. They have to maintain current applications (for instance, payroll programs and accounting programs must reflect new city taxes and different depreciation schedules) while at the same time develop new computerized systems needed by the rest of the organization. Requests for processing often exceed the available resources, so rationing occurs, and someone must set priorities. Those frustrated individuals with low-priority projects often turn to personal computers instead, and may find that computer-related tasks now take less time, cost less, and are more accurate. And when analyses are better, so are decisions.

Using a tested software package on a microcomputer is usually a productive experience, but limits can be quickly reached on transaction volume and data storage capacity. For example, if the software package is designed to handle 250 sales transactions a day, there will be problems when peak days reach 300 transactions. Or if the package was designed for a church with 500 members, storage difficulties will result when the membership exceeds 500. When user expectations go beyond the data-processing and storage capacities of the microcomputer, we're right back where we started: we have to go to the data-processing department for help and are rebuffed or forced to stand in line.

The data-processing department in most organizations typically has a big *systems and programming* backlog, which means that it is responsible both for the design of information systems and the writing of programs to implement them. Therefore, the department is not likely to be interested in applications where it does not have complete authority over operating procedures and data integrity. Also, department programmers might have to be retrained to write programs for microcomputers, which are often highly interactive. Achieving a satisfactory human/machine interface takes training and experience.

Let's assume a different scenario. Rather than programming help, let's say you need access to company files from your own computer. Entering the data yourself would be an expensive duplication of what data processing already has in its central mainframe. The hardware and software to download it directly to you from data processing could be expensive, and there is also the question of corporate approval to use the data locally. Destruction of files and alteration of information are among the concerns that data-processing people will mention.[1]

Some twenty years ago, an aerospace firm automated much of the production information in its aircraft factory. There were terminals scattered around the plant for input and output. A worker on the swing shift assumed the nearby terminal could be used as a typewriter, and during his lunch break, he typed a letter to his brother. By accidentally keying in the code that provided access to an important file, his subsequent typing created many errors in that file. It is such random romps through the company files by personal computer users that data-processing people and management fear most.

INTEGRATION

A reasonably typical office automation situation might include:

- a mainframe computer with time-sharing and perhaps some distributed data processing
- quite a few individual word processors
- a growing personal computer population of stand-alone micro-computers
- many telephones, all controlled by a telephone switch (Private Branch Exchange)

People use the spoken word, the written word, numbers, and pictures to assimilate information and to communicate it to other people. These forms are integrated and used simultaneously, as when you discuss a written report by phone and analyze one of its graphs for your listener. In a similar manner, we are slowly and imperfectly integrating the technologies of data processing, word processing, image processing (i.e., microfilm), audio processing (voice), and networking.

The IBM Personal System/2 illustrates how hardware is moving toward integration. It can act as a host to other personal computers, perhaps storing a data base on its hard disk which other computers can access for their own use. We can also connect it to a local area network, thereby sharing its resources (e.g., peripherals, software, information) with other devices located on the same premises (e.g., office building, plant, or campus).

The market for personal computer software also demonstrates this trend toward integration. Early **spreadsheet** programs like VisiCalc provided an electronic worksheet divided into rows and columns that helped to organize and present business data, but did little else. Later spreadsheet software like Lotus 1-2-3 performed several functions, including business graphics and data base management. The latest generation of programs like Top-View, Framework, and Microsoft Windows allows the user to integrate individual packages. Therefore, Lotus 1-2-3, Microsoft Word, and dBase III Plus might be used together, letting you create spreadsheet information with Lotus, store that information in dBase III, and then extract records to include in a report you've produced with Word. The multitasking of the Personal System/2 is a more recent expression of this trend.

IBM's acquisition of Rolm, a manufacturer of telephone equipment, suggests that the leader in the computer industry has committed itself to the idea of integration. There seems little doubt that we are moving, although slowly and imperfectly, toward an integration of computers and communications.

BUSINESS TELECOMMUNICATIONS

A few years ago, the typical mainframe cost about $80 per hour to run. Today, desktop computers cost about 60¢ per hour.[2] This drop in price applies not only to computers but to other hardware as well.

In the 1950s it cost the Bell System about $60 for an average mile of long-haul circuitry.[3] This dropped to around $25 by the mid-1960s. Later, the coaxial cable systems cost something like two dollars per circuit mile. The falling costs of business telecommunications systems mean that applications which were once impractical can become feasible almost overnight. Since personnel-related costs such as travel and mail delivery are growing, business telecommunications become more and more attractive.

In a few years, there will be few institutions without computer-based communications. Networks will soon be able to integrate all the various forms of information an organization needs: data, text, images, voice, and graphics. These networks will become a business within a business, helping to improve productivity.

People and computers are partners in the processing of data and the distribution of information to help solve business and scientific problems. As illustrated by Figure 1–2, the computer half of this partnership is composed of hardware, software, and **firmware:** software instructions that have been more or less permanently placed into computer memory. The business telecommunications environment is a complex mixture of products that are often acquired from different vendors to operate at different network locations, and common standards are needed so that all of these heterogeneous elements can communicate. These standards are being developed by the *International Standards Organization (ISO),* a voluntary organization composed of the national standards committees of each participating country.

ISO's standards are embodied in its reference model for *Open Systems Interconnection* (OSI). Network complexity is managed by grouping similar functions into seven layers, each of which allows interaction with adjacent layers (see Figure 1–3). Each layer is relatively self-contained, with clearly defined interfaces, so a change made in one layer need not impact elsewhere. This structure of different layers for different functions and ser-

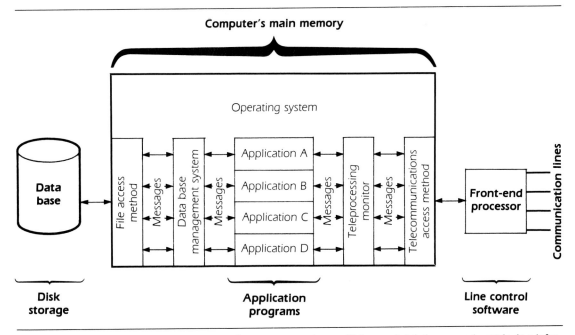

FIGURE 1–2 *Messages may include creating new records, modifying and deleting records, retrieving information from the data base, controlling other devices (e.g., plant machinery), and routing electronic mail.*

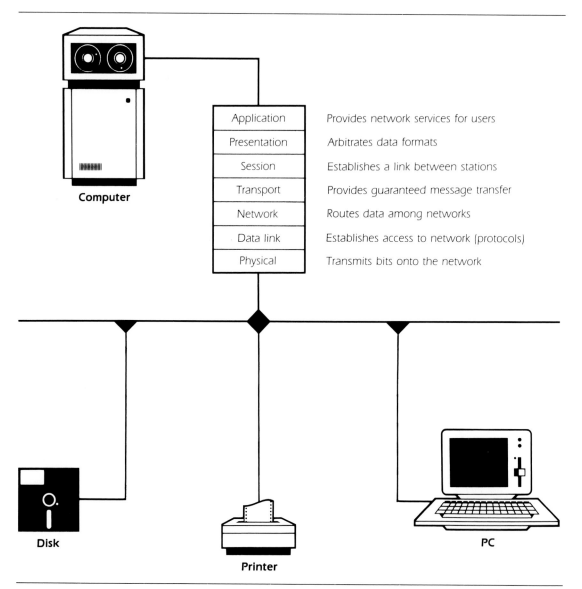

Application	Provides network services for users
Presentation	Arbitrates data formats
Session	Establishes a link between stations
Transport	Provides guaranteed message transfer
Network	Routes data among networks
Data link	Establishes access to network (protocols)
Physical	Transmits bits onto the network

Computer

Disk

Printer

PC

FIGURE 1–3 The OSI communication model spells out the seven basic tasks that must be accomplished before communications can take place on a network. Messages pass from the sender through each of the seven layers and onto the network. They then travel back up through the layers to the recipient.

vices tends to move telecommunications away from incompatible, one-time only solutions and toward generic solutions, independent of any particular hardware or software products. The seven-layer model enables machines from different vendors to communicate, even if they are located on different continents (see Figure 1–4). It is this structure that is the basis of network architecture, and it is typical of the material we will cover.

——— THIS BOOK ———————————————

The first step in our mastery of business telecommunications is to take a general look at how a terminal talks to a computer over the public telephone system (Chapter 2). Chapter 3 then discusses how we are moving toward digital telecommunications as we replace and expand our existing telephone system.

More business telecommunications concepts are discussed in Chapter 4, including public packet-switching networks. This is followed by a

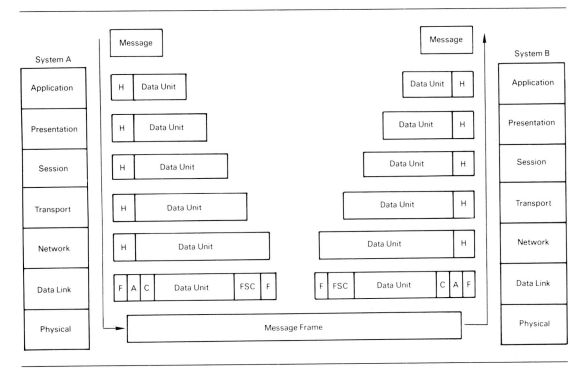

FIGURE 1–4 In this illustration, a data message is transmitted from an application program in System A to an application program in System B. As it is sent, each communication layer manages or alters the message in some way and adds information for its counterpart in the form of a message header (H). Each layer's header, when added to the original message, forms part of the data unit. In the Data Link layer, flag (F), address (A), control (C), and checksum (FSC) bits are added. These bits complete the message frame that is transmitted over the physical layer of the network to System B. As the data unit passes through System B, each layer removes and reads the header intended for it, performs the appropriate data conversion, and passes the data unit on to the layer above it.

Wang Laboratories, **A Primer on Computer Networks** (August, 1982), No. 777-4004, 61. Courtesy Wang Laboratories.

consideration of some software concepts useful in telecommunications (Chapter 5).

Chapter 6 discusses protocols: how a logical communication channel is established and how transmission then occurs. Having reviewed software and protocols, we then tackle network architecture, and the International Standards Organization's seven layers are covered in Chapter 7. Next, we turn to local area networks in Chapter 8. Standards, regulation, and divestiture are covered in Chapter 9; comprehension of the regulatory environment is essential if we are to understand the telecommunications market.

The book ends with a discussion of telecommunications and management (Chapter 10), network design (Chapter 11), and errors and security (Chapter 12). The glossary at the back of the book defines many of the terms used.

Without giving it much thought, almost all of us have been involved with telecommunications simply by using the telephone system which has existed for over 100 years. As long as we make the desired connection when we dial a number or push the buttons, we give little or no thought as to what the system is about or how it operates. Interestingly, though, it was the telephone industry that paved the way for today's telecommunications expansion and the multitude of products and services we now have to help us. It seems appropriate to start learning about business telecommunications by looking at the telephone network, which is what we do in Chapter 2.

THE CASE STUDIES

About a century ago Joseph Wharton, a financier and iron manufacturer, noted that universities had done little to prepare anyone "for the actual duties of life" except lawyers, doctors, and clergymen, and that commercial education was left to "private colleges" below the university level. He thereupon endowed the Wharton School of Finance and Economy at the University of Pennsylvania in order to offer a specialized education "in the principles underlying successful business management."

Since the Wharton School started with thirteen students in 1881, colleges of business administration have faced a quandary: business management is best learned by the actual experience of management in the commercial world, and there is no substitute for such experience. If one accepts this premise, one must then ask: What can schools of business do to help people learn to manage?

Dean Wallace B. Donham of the Harvard Graduate School of Business Administration provided one answer during the 1920s. Heavily investing the school's financial and intellectual resources, he developed case materials with which to stimulate actual business decision making. This use of business case studies is often compared to training physicians with actual medical cases: the intern must ascertain what is fact and what is opinion and then diagnose and finally prescribe. Management trainees go through a roughly similar process with management cases.

The case study approach has proven itself time and time again, and today, many U.S. and foreign universities employ case studies. Analyzing cases is one of the few ways in which students can simulate actual management experience in an academic setting. In addition, case analysis automatically reviews previous material by bringing it to bear on real problems. Defining and solving these problems helps to train students in decision making.

Cases are collections of raw information about a business. Deliberately unstructured, the problem is seldom defined. Individuals may give their appraisals of the situation, but these opinions do not necessarily reflect good business judgment.

Sometimes names and locations are altered, but otherwise the case is an accurate report about the state of a real business, as well as an accumulation of information about a particular management problem much as the executives of that firm encountered it. Cases should not be construed as representing either good or bad business practices but are simply raw material to help the student become a more proficient manager.

You will consider two cases in this book: the Brisfield Company, which begins on p. 13, and Clark Fibers, Inc., which is at the end of Chapter 8. You will encounter questions about the Brisfield Company at the end of most of the chapters in this book. In contrast, the Clark Fibers case is confined to the chapter on local area networks.

Read the Brisfield Company case next. After you do this, prepare a brief summary of the company's business telecommunications development. The Brisfield Company is a fictional organization.

SUMMARY

1. Telecommunications is now the premier form of information technology.

2. Computers and communications are both essential parts of automation.

3. We are slowly and imperfectly integrating the two technologies.

4. The falling costs of computers and telecommunications mean that applications which were once too costly can become feasible almost overnight.

5. Telecommunications is moving away from incompatible, one-time-only solutions and toward the generic solutions embraced by the International Standards Organization.

REVIEW QUESTIONS

1. "Telecommunications is now the premier form of information technology." Comment.

2. "Computers and communications are both essential parts of automation. As we connect them, we will cut costs, raise quality, and improve productivity." Explain.

3. Trace how the technology has became cheaper, more powerful, and smaller. What are the implications?

4. Define: mainframe, local area network, firmware, and International Standards Organization (ISO).

5. Describe how we are slowly and imperfectly integrating computers and communications.

6. Explain Figure 1–2. What are some examples of messages?

7. Using Figures 1–2 and 1–3, describe how the International Standards Organization model allows two machines to communicate.

ENDNOTES

1. Frank Greenwood and Mary M. Greenwood, "Power to the People," *Journal of Systems Management*, December 1981, 10.
2. NCR, *Consultants Communique*, Jan./Feb. 1985, 4.
3. W. C. House, *Electronic Communications Systems* (New York: Petrocelli, 1980), 158.

THE BRISFIELD COMPANY

The Brisfield Company was founded in 1909 by two brothers, John and Joseph Spooner, who had unsuccessfully attempted a similar venture with another brother, Gideon. Their first endeavor had strained the allegiance of the two younger brothers, and thus Gideon, the eldest brother, was left to operate the start-up enterprise in Amagonia, Connecticut. John and Joseph, unhappy with the "bossiness" of Gideon (who was a college professor during the week and came up to Amagonia on weekends), decided to break away and organize their own firm. They found a reasonably priced three-story mill building in the eastern Massachusetts town of Brisfield which would allow them to escape the domination of their older brother and yet remain close enough to Amagonia to maintain their Spooner family ties.

After moving up to Massachusetts, the business (which in those early days was making very simple industrial widgets, in direct competition with Gideon's Mansboro Company) was greatly stimulated by large orders from a nearby naval base. The Brisfield widgets were found to be very useful in the guidance control used by torpedoes during World War I. This allowed the company to grow under the two brothers' careful but enthusiastic guidance, and to acquire important new technology. The firm was further stimulated by World War II and again altered its product lines to assist in the war effort. Growth over this long period of time averaged about 20% annually, and the firm added facilities for building, selling, and serving their widgets on a worldwide scale.

John and Joseph Spooner, the founders of the company, each had large families. Their children all grew up in Brisfield, and the brothers agreed that each would train his own brood of young Spooners in the basics of the business. This they did. The second-generation Spooners all worked at the company and learned to appreciate the functions and contributions of employees at all levels of the operation. In each family there was one son and several daughters; the daughters eventually married or moved away from Brisfield, but the two sons dedicated their careers to the family company. In the years after World War II, the elder Spooners retired and the next generation took over the job of managing the firm.

By the 1950s, the Brisfield Company had manufacturing plants in Brisfield, Spanagua, and Vixenmead, Massachusetts, as well as small assembly installations in Atlanta, Chicago, Houston, Pittsburgh, and San Diego. They had also become an international corporation and owned manufacturing facilities and sales offices all over the world. The largest of these industrial holdings outside of the United States included manufacturing plants in Halifax, Nova Scotia; Queensberth, Holland, Pinkcroft, England; Blossomglen, Australia; and Mexico City.

Doing business in the international marketplace with customers of all sizes (including the largest companies in the world) required Brisfield to make serious commitments to good communications capabilities and to upgrade them as the technology allowed.

Brisfield's widgets came in many varieties and the company had major markets in the following key industries:

- chemicals
- petrochemicals
- oil and gas
- power
- pulp and paper
- food
- metals
- mining

Brisfield products were continually evolving as new technology provided the needs and the means, and the product line grew larger and more complex. Earlier widgets were "standalones" without interconnections and communication linkages, but the development of modern electronics and communication technology greatly influenced widget development. Modern models were often collections of interconnected units with complex electronic functions that could communicate with both specialized and general-purpose computers. Likewise, because of safety requirements in some of the volatile (and potentially combustible) environments where widgets might be mounted, pneumatics as well as electronics had to be used for operation and signal transmission. This required additional hardware to integrate the electronic and pneumatic information, and widget signal converters became part of the product line.

By the 1970s, Brisfield's annual sales volume was above $500 million. The company had started to use computers as soon as the technology was available, since its technical staff included recent graduates of the best technical universities in the country. The two second-generation Spooners running the company during this period, Eli and Jesse, were both graduates of MIT in nearby Cambridge, Massachusetts.

Remote Computing

In the mid-1960s Brisfield engineers were discovering that software for specialized computer applications was available if they could get approval to spend money on a new technology called a *service bureau* or *remote time-sharing*. Since the cost of these service bureaus was much cheaper than buying the software or developing it themselves, and since the service bureau computers were much larger and faster than those at Brisfield (an IBM 1130 and an IBM 650), the engineers began to bring their data and work to computing centers in nearby Cambridge, Boston, and Waltham. As they interacted with the staff at some of these centers, they were introduced to a potentially better idea: *remote access computing*. This took several forms, all of which at least saved them the time, inconvenience, and cost of traveling to the computer sites. In some instances, the engineers and scientists transmitted *job decks* (programs and data in punch card form)

via a remote card reader hooked onto a Bell Telephone Company *modem* (a device for transmitting data over voice telephone lines which closely resembles a normal desktop phone). The data from the punch cards was transmitted to a receiving card-punching device also hooked onto a modem. Once the job deck (with appropriate header cards identifying the client) was received, it would be queued onto the service bureau computer, which would then run the jobs and produce printed output. This would usually be mailed back to Brisfield.

One of the earliest applications for this new technology was to automate many of the technical calculations required when sizing widgets to customer specifications. These calculations used to be done manually, and errors had caused serious malfunctions in customers' plants. Another consequence of this time-consuming work was a tendency by the engineers to "play it safe" and to provide bigger widgets than were really necessary. The unfortunate aftermath was that customers bought larger widgets than necessary at a higher cost, and sometimes business was lost to other vendors with more precisely sized products. The new approach saved time and improved accuracy, and thus improved the business. Besides, it was exciting for the engineers and scientists, who felt they were on the leading edge of their professions.

The more Brisfield used these facilities, the more it realized how much could be gained by adding to the applications on these computers. This began a period of rapid growth in the use of communications and computers for product development and engineering support, as well as for sales order engineering and, eventually, business management functions.

Competition among the service bureaus selling this type of computing service soon began to improve the delivery of the service to users such as Brisfield. Before long, a user could transmit a job directly to the computer, where it would be stored on a direct access storage medium which could be accessed from a remotely connected terminal. Appropriate line-editing software allowed the user to "look" at what was stored in the file by printing it on the remote terminal at Brisfield. Even better, the user could now make changes to the job deck and resubmit it without having to retransmit it over telephone lines, which greatly improved productivity.

Later, the development of *cathode ray tubes,* interactive computer operating systems, faster and more reliable modems, faster printer terminals, and on-line program debugging tools all spurred the use of such remote time-sharing systems.

Inventory Control

One of the first business applications of data transmission over telephone lines at Brisfield involved the widget inventory control system. Brisfield had five field warehouses where popular models from their product line were kept in inventory for rapid delivery to customers. The locations of these field warehouses, which also served as repair shops and supply sources for spare parts, were in metropolitan areas where there were major concentrations of industrial customers. The five cities were San Diego, Chicago, Atlanta, Pittsburgh, and Houston. The northeast region was serviced out of the central finished widget warehouse, located in Vixenmead, Massachusetts, a town just south of Brisfield.

The purpose of the application was to make more accurate inventory information available to planners and order clerks at the central warehouse. As orders were routed to the central warehouse for processing, it was critical to know precisely how many of which models were in each field warehouse. By dispatching the product from the closest point, delivery time as well as transportation cost would be minimized. It was thus very important to have an accurate, centralized record of what was in inventory at all times.

In addition, the field warehouse inventories were supplied from the central warehouse, which ordered its replacements from the manufacturing plant. Lead time from manufacturing was four to eight weeks, depending on the model. Widgets could be packed and shipped from the central warehouse to any of the field warehouses and, depending on the time of day, be delivered to the customer the next business day in an emergency. However, it would be more cost effective (and quicker) for the field warehouse to provide the widgets from its own inventory.

Likewise, in the event that some lower-volume item was needed, perhaps one not available in the central warehouse, it would be useful to know if it were in stock in some other field location. That information would allow the product to be shipped immediately, and would save the four-to-eight-week manufacturing time.

In order for the inventory records to be accurately maintained, all remote inventory transactions would need to be transmitted to one central place. It was clear that the movement of widgets out of the field warehouses, which usually took place in small quantities and at random times, would create the biggest challenge, since this information would have to be carefully recorded and transmitted in a timely fashion to a central site for periodic updating of the inventory status. Inventory shipments from manufacturing to the central warehouse and field warehouses and shipments from the central warehouse to field warehouses were usually planned well in advance and involved larger quantities, so recording these transactions was less of a challenge. Trans-shipments between field warehouses were relatively rare.

The final solution was a star network of IBM 1001 data transmission terminals, connected to the public telephone system via Bell modems and communicating to a card-punching receiver at the central warehouse, which allowed real-time transmission of transactions at any time of day. The receiving card punch had to be made ready to receive and be loaded with blank cards by an operator but could receive in an unattended mode. The normal procedure was to begin each week with a printed master list showing the status of all widgets on hand and on order by warehouse location, including the central warehouse.

All physical movement of widgets was batched and recorded on transaction cards at the central receiving terminal. These transaction cards were periodically gathered by a supervisor, and the start-of-week inventory report was marked up by hand, reflecting the actual inventory changes occurring during the week. Thus, the operator at the central warehouse always knew the precise status of inventory throughout the whole system, and anyone could get this information by telephone. The field warehouses likewise marked up inventory reports to reflect weekly changes in their own stock. On Friday evenings, all accumulated transactions were batched and input into the program which updated the central computer status file, and the new start-of-week master lists were created and distributed. By Monday morning, a fresh, accurate copy was available at all the warehouses.

To facilitate these field-based inventory reports, all widgets shipped to the central or field warehouses were accompanied by a prepunched card which recorded all of the pertinent data about the unit. This *turnaround document* provided most of the transaction information that was transmitted via the IBM 1001 terminal, with the operator adding a few simple codes to identify the model number, serial number of the product, warehouse code, and any other data fields. The receiver was programmed to automatically enter the date on the output card.

Product Planning

During this era at Brisfield company engineers and product developers conceived a brilliant new concept for their product line. Traditional widgets, even when they were interconnected to perform complex functions, were built with their own metal casings. This design dated from the earliest days when the widgets were almost always field-mounted stand-alone units.

As Brisfield became more sophisticated in applying widgets to complex processes and operations, a larger proportion of their products either used or interconnected with electronic devices. This increased the need for transferring information back and forth between different widgets, which in turn became a system for providing the customer with operations guidance and regulation functions.

A key element in the purchasing decisions made by Brisfield's customers, a rational group of intelligent businesspersons, was the cost of owning any of the competitive systems of widgets that could be used for operations guidance and regulation of their plants. A very significant factor in this choice among competing vendors—in addition to the cost of acquisition—was the cost of supporting the system and keeping it in operating condition throughout its useful life. Clearly, since the operation of whole plants might depend on such systems, the system had to be kept in working order. To do this at a reasonable cost meant that the system should incorporate ease of maintenance into its design, and not just as a cosmetic attribute. Thus, *lifetime cost,* not simply acquisition cost, was a decisive factor when procuring widgets and widget systems for operations guidance and regulation.

One of Brisfield's most creative marketing and product strategists, Foster Gifford, anticipated that the marketplace would respond favorably to these new concepts and proposed that the company create a family of widget systems which, as much as possible, would eliminate individual widget packaging (and the concomitant costs). Instead, the new "Unique–50" product line would utilize a new flexible multifunction widget packaging, which would simplify service and maintenance.

Different widget models could be housed in *nests,* which might hold as many as ten widgets; multiple nests could be housed in *racks;* and multiple racks might be bolted together in rack configurations to match the customer's work area. Since the individual widgets were made in standard sizes, the nests could be configured with different combinations of widgets. Nest locations could even be left empty as a provision for future expansion.

The advantages were many: a design flexible enough to meet the unique requirements of customers; lower acquisition cost; lower maintenance and support costs; and a great improvement in reliability. The new approach also made installation and alteration of the system much simpler. By housing the modules in nests and racks with built-in connections, servicing was greatly simplified. Key dimensions and electrical attributes were standardized to the marketplace, giving this product line an additional advantage over the competition.

The Worksheet

One major complication created by this new product concept was the problem of configuring a system to meet the client's needs. Since the product line offered a comprehensive series of operations guidance and regulation functions, the possible combinations and permutations of these functions in any control system was practically unlimited. Finally, since so much of the packaging and configuration of the system was based on customer choice, Brisfield's sales engineering organization (a worldwide force of more than 350 experienced professionals) had its work cut out for it.

With older products, quoting a job for a customer required very little effort on the part of a sales engineer. "Sizing" a widget was a simple computationally based decision made by home office experts who had computer support. Likewise, all field sales offices of any size had resident support engineers to whom any complex price analyses or application design choices might be referred. The field sales force was further backed up by product line specialists and industry specialists back at the home office in Brisfield. The new product line, however, was going to require sales engineers to do a lot more calculations than ever before. This was recognized long before the research and development effort to create the new Unique–50 product line was completed.

A solution to this problem was delivered along with the rest of the product line specifications, preproduction drawings, tools, and procedures. This solution had been worked out by two of Brisfield's veteran employees: an assistant to the order entry manager, and a senior member of the home office's product sales department who had specialized in the new product line. The solution was a fold-out worksheet which let the sales engineer fill in the basic choices and functional selections needed to satisfy the customer's requirements for operations guidance and plant regulation. From these initial answers, the worksheet guided the user along a series of data "look-up" tables which eventually derived answers to the following questions:

1. **How much power** (of each of four different types) would the total configuration require?
2. **How many power supplies** (by model number) must be included on the quotation?
3. **How many nest spaces** (of each of four different types) were required?
4. **How many nests** (by model number) should be included on the quotation?
5. **How many racks** (by model number) should be included and what configuration should they take?
6. **How many cables** (by model number) should be quoted and priced?

The worksheet was accordion-folded and, when extended to its full length, was two-and-a-half feet long. Adding insult to injury (in the eyes of the field sales organization), the form was printed on both sides and was covered with a maze of overlapping and intersecting arrows.

At this point, the bulk of Brisfield's field sales engineers were not used to selling systems for operations guidance and regulation. They had grown up selling individual widgets or, sometimes, a collection of different widget models interconnected to perform some complex function. Not only were they required to learn a whole new product line, but in the meantime they had to continue to produce bookings to deal with the company's backlog.

One of the more aggressive field sales managers produced data showing that, in order to sell his quota of Unique–50 widget systems, he would immediately need 40% more salespeople or five new sales engineers. Numerous howls of complaint were raised and it was soon evident that something needed to be done. The sales management approached Brisfield's top management with a top-priority request for computer assistance.

CyberQuote

By this time, Brisfield Corporation's information services group had coalesced into three loosely connected subgroups:

1. a mainframe group, running COBOL-based batch-processing applications on the corporation's IBM mainframe computer,

2. a remote-processing group, supporting the newer time-sharing interactive computing technology,

3. the corporate telecommunications group, working closely with the remote-processing group and also supporting the company's telephone and internal mail facilities.

The sales management's request for help landed — with the highest priority — in the laps of the remote-processing group.

During this period some of the big time-sharing companies — National CSS (which later became Dun & Bradstreet Computing Services), Tymshare, and General Electric Time-sharing — were attempting to sell Brisfield their services. For a variety of reasons, some specialized computing services were purchased from smaller vendors, but the bulk of Brisfield's time-sharing was bought from National CSS and, to a lesser extent, GE Time-sharing.

Both National CSS and GE Time-sharing had gone international by this time, and both offered their clients the opportunity to make a local phone call from locales around the world and instantly connect into their central computer in Norwalk/Stamford, Connecticut, or Cleveland, Ohio, respectively.

Fred Richardson, Brisfield's time-sharing manager, believed that the Unique–50 sales dilemma was a tailor-made opportunity to show Brisfield the benefits of the new communications and computing capabilities. His enthusiasm was shared by his colleague, Monk Jenney, manager of corporate telecommunications, who was having difficulty persuading top management to invest in the emerging computer and communications technology.

A prototype program was developed to recreate the complex Unique–50 worksheet on the computer, which would interactively ask questions, accept and validate answers, and, once all required inputs had been entered, determine all the power supply, nest, rack, and cable requirements and print them out on the user's terminal. This prototype was used to introduce the concept of interactive computing to the sales specialists in order to determine which design parameters were missing, which needed changing, and which hit a resonant note with prospective users.

Feedback from the users and their managers resulted in some refinements. Added to the system requirements was the capability to create "green sheets" for customer quotations, item-by-item price explanations which included the current list price in U.S. dollars, the currency conversion factor (if any), discounts, extensions by quantity, and the final price. The users also specified the layout of this output report, as well as the layout of the quotation itself, which they wanted to use as an attachment to the submission letter prepared by the account sales engineer. This letter would be generated at the sales office responsible for dealing directly with the client.

The users also wanted to be able to revise a quote, to prepare different versions of the same quote, and to be guided through their answers to the queries. They were very adamant that clearly impossible or erroneous answers be rejected by the system, which meant that the system needed to thoroughly screen all responses.

Fred and his staff selected GE Time-sharing as the primary time-sharing service for this application since they had the largest worldwide network. Programs were developed for three separate levels of editing and validation:

- *format checking* — numeric, alpha, in specific locations

- *content checking* — responses checked against a list, or against upper or lower limits, etc.

- *context checking* — checking for acceptable combinations and logical conditions (more difficult to specify and implement)

The system was designed to provide the user with a lot of help and specific suggestions as he or she interactively responded to requests for information about the customer's application requirements.

Someone named the application *CyberQuote*, and it caught on. It worked like this: once a sales engineer had successfully satisfied the computer by providing all the right answers, a facsimile of the finished worksheet was printed at his or her terminal and stored in the time-sharing computer.

At this point, the engineer knew that the configuration was technically and logically correct, even though the total cost had not been calculated or all the line items determined. He or she could optionally request the actual printed quotation, the green sheets, or both, and depending on the urgency choose to do it interactively (in other words, immediately), in "background" or deferred batch, or in overnight batch. The engineer could also specify how many copies to print out.

After the system had been introduced on a limited basis into the national sales organization, a user review pointed out some serious problems: the time-sharing costs per sales office were far beyond what was deemed reasonable, the printing quality was inconsistent and too often unacceptable, and sales engineers were spending too much of their time at computer terminals and not enough time selling widgets.

Fred Richardson and Monk Jenney conferred with their GE Time-sharing technical representative, Joan von Braun, looking for ways to improve the system to the satisfaction of field sales. The interactive sessions during which answers were provided to the program were averaging less than twelve minutes per order, but some of the computations and printing the reports were taking more than five times as long. In addition, the computations were chewing up computer time at a rapid rate, further adding to the time-sharing charges.

While not all field sales offices were using the CyberQuote, enough were to determine the average weekly usage they could expect from the 42 sales offices in the United States and the 35 sales offices in the rest of the world:

NO. OF OFFICES	AV. QUOTES/WK	COMMENTS
15	7	Small field offices
23	16	Medium/small field offices
37	38	Medium field offices
2	55	Large field offices
5	35	National home offices

The conference with Joan and Monk convinced Fred that he should look further into the technical feasibility of communicating directly between the GE Time-sharing computer in Cleveland and the Brisfield IBM 360/50. Monk was also enthusiastic about this possibility, and after determining that he had the appropriate hardware and testing the software provided by GE with Joan's help, he asserted that it was possible to transfer files from GE to Brisfield on a deferred batch basis at a cost of less than 50¢ per thousand bytes (characters). The average number of variable characters in an unformatted worksheet was 496; in a fully formatted worksheet, the number of characters was 1978; in the average green sheet there were 23,000 characters; and in the average quote there were 32,000 characters.

Private Network

In 1977, corporate requirements for telephone, telex, and data services were analyzed and a cost-justified, satellite-based private network was installed. Paradyne equipment was chosen on a three-year lease at a cost of $164,000 per year; the lease was renewed once in 1980 and was up for renewal again in January, 1983.

One of the major data services supplied at this time was computer-aided design on a series of DEC minicomputers, a system which allowed engineers in different parts of the world access to common tools and information. The primary locations for these engineers were Pinkcroft, England; Queensberth, the Netherlands; and Brisfield, Massachusetts, although the DEC computers were located in Brisfield. This system permitted sharing of work without the delays ordinarily caused by older communications technology, allowed Brisfield to improve its cycle time for bringing new products to market, and enriched the quality of the designs themselves.

In order to improve the firm's ability to fill the complex, multisourced orders often placed by major international firms, Brisfield installed a communications-based *major projects reporting system* in its international network of sales offices. Sales and custom engineering systems also centralized their order-filling information in a computer back at Brisfield and communicated via the network.

When the Paradyne lease came up for renewal again, the international telecommunications line was at full capacity, and voice calls had to be diverted back to the public telephone network. The annualized cost of diverting approximately 450 calls per month was $72,000. In order to meet the levels of data transfer required by Brisfield's European Technology Organization and the major projects reporting system, $32,000 per year was being spent on the public IPSS packet switching network and Telenet. The breakdown of these costs was as follows:

	COST/HR		HRS/MONTH		MONTHS		TERMINALS		$/YR
Technology	$15.00	×	105	×	12	×	1	=	$18,900
Sales	15.00	×	25	×	12	×	1	=	4,500
Fixed Network costs									8,600
Total									$32,000

The projected rise in costs resulting from increased use was estimated to be $129,000 annually, assuming no change of network:

	COST/HR		HRS/MONTH		MONTHS		TERMINALS		$/YR
Pinkcroft									
Technology	$15.00	×	105	×	12	×	2	=	$37,800
Systems	15.00	×	25	×	12	×	1	=	18,900
Queensberth									
Systems	$28.00	×	105	×	12	×	2	=	$70,560
Fixed Network costs									$1,740
Total									$129,000

These estimates were based on the use of the Dutch DBAS network, which, based on previous experience, would be of marginal quality. This mode of expansion would therefore not meet the level of service required.

Monk Jenney and his capable staff proposed two alternatives to the current system:

1. *A separate network for the European Technology Organization.* This would meet minimum technology requirements and allow the company to regain voice communications savings of $72,000 per year. The 1977 Paradyne equipment lease would be renewed or the apparatus would be purchased outright. Capital expenditures of $50,000 for additional equipment would be required. Those who did not want this solution pointed out its deficiencies:
   - It did not support other corporate needs.
   - Its capacity for growth was limited.
   - It would not allow the combined efficiency of dual lines.
   - Additional technical support would be required
   - Packet usage from Queensberth would most likely require the diversion of voice traffic again in the future.

2. *Replace the Paradyne equipment with two leased cable lines* connecting Infotron 790 Telecommunications Processors in Brisfield,

Pinkcroft, and Queensberth. This solution would allow the company to recover voice communications costs of $72,000 per year, eliminate packet Network charges of $32,000 per year, and avoid expected 1983 increases in Packet Network charges of $129,000, as well as even higher costs in future years in demand for this capacity (growth was projected at 25% per year). The cable lines would provide transparent interfaces (interfaces invisible to users) to multiple networks with different protocols, codes, and speeds.

Memorandum (Draft)

To meet the immediate needs of the European Technology Organization and those of other Brisfield Corporate groups, it is recommended that the existing satellite link and the Paradyne equipment be replaced with two leased cable lines connecting Infotron 790 Telecom processors in Brisfield, Massachusetts, Queensberth, the Netherlands, and Pinkcroft, England. This solution would allow us to recover voice savings of $72,000 per year, eliminate packet network charges of $32,000 per year, and allow us to avoid the expected packet costs of $129,000 for 1983 as well as future costs estimated to be growing at a rate of 25% per year. In addition, the network would provide the company with transparent interfaces to multiple networks with many current protocols, codes, and speeds.

Data transfer between nodes would operate with two combined 9.6 kilobit lines at 19.2 kilobits per second for 18 hours per day (1:00 P.M. to 7:00 A.M. BRISMASS time) and with one line at 9.6 kilobits per second during the remaining six hours when the other line is used for voice.

This proposed configuration will support up to 21 interactive terminals at 1200 bits per second 24 hours per day, with increased capacity for batch data transfer 18 hours per day.

Additional terminals or lines could be added in a modular fashion to this "backbone" network as required, to a maximum of eight trunk lines at 72 kilobits per second and 448 terminal ports up to 9.6 kilobits per second. Costs for the system would be as follows:

	EXPENSE/ YR	CAPITAL	START-UP
Leased network costs	$271,592	——	$2,962
Installation	——	$1,050	2,000
Capital equipment	——	131,741	——
Total	$271,592	$132,791	$4,962

Implementation could begin in April, 1983, with the addition of one international cable line and Infotron equipment at Brisfield, Pinkcroft, and Queensberth. This will immediately improve the service to the technology organization. In June, 1983, the existing international satellite line could be replaced with a cable circuit and used on a switched basis to provide additional data capacity alternately with voice service. All existing telex services could be transferred to the new circuit and upgraded at that time to two 1200 bit-per-second access lines.

2 DATA TRANSMISSION OVER PHONE LINES

AFTER READING THIS CHAPTER, YOU WILL UNDERSTAND THAT:

- Our telephone network has four basic parts: telephones, local loops, switching centers, and the transmission network.
- A terminal talks to a computer over the analog public telephone system using codes and modems.
- Two widely used codes are ASCII and EBCDIC.
- Digital signals are discontinuous, and modems modify digital data so it can travel over analog (voice) circuits via continuously varying electrical signals.

————THE TELEPHONE SYSTEM————

To understand business telecommunications, we need to first review how our telephone system works. Our voice telephone network in the U.S. is old as well as huge: patented in 1876, it now includes over 170 million telephones. Figure 2–1 shows the system's basic parts, which are:

1. **Telephones.** The telephone mouthpiece converts speech into electromagnetic impulses that are transmitted over the telephone lines and reconstructed into speech by the receiver at the other end. The transmitter and receiver are usually contained in the handset, and the bell and switch hook (the bar or buttons under the handset) are located in the base. Calls are placed using a rotary dial or push buttons.

2. **Local loops.** Local loops (also called feeder cables, drop wires, or distribution cables) connect telephone subscriber to the phone company's local offices. A local loop, through which all telephone services are provided, is usually a pair of copper wires one or two miles long. The loop operates on low-voltage direct current supplied by rectifiers (which convert alternating current to direct current) and backed up by batteries and diesel generators at the local office.

3. **Switching centers.** Switching is done by a hierarchy of central offices (see Figure 2–2) which select the transmission paths to complete calls. Each office handles the phones in its area, and higher-level offices handle larger areas and more telephones. For instance, a local exchange might have 10,000 lines, while a regional center might handle over ten million. The electronic switching system of a local office illustrates how this works (see Figure 2–3). The trunk or long-distance lines and the subscribers' local lines all enter the electronic switching network **(A),** and the computer **(B)** determines the path of a call through the transmission network (different pathways may connect the same two callers at different times). The scanner **(C)** monitors line activity, such as when a phone is picked up or put down and when various incoming calls from other switching offices comes down the trunk lines. The signaling equipment **(D)** generates dial tones, busy signals, and any other signals that have to be sent out. The computer collects information for billing and network analysis by the phone company **(E).** When

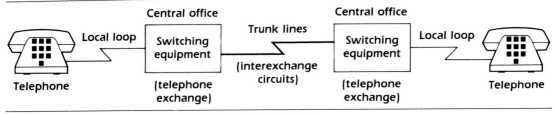

FIGURE 2–1 Basic voice network.

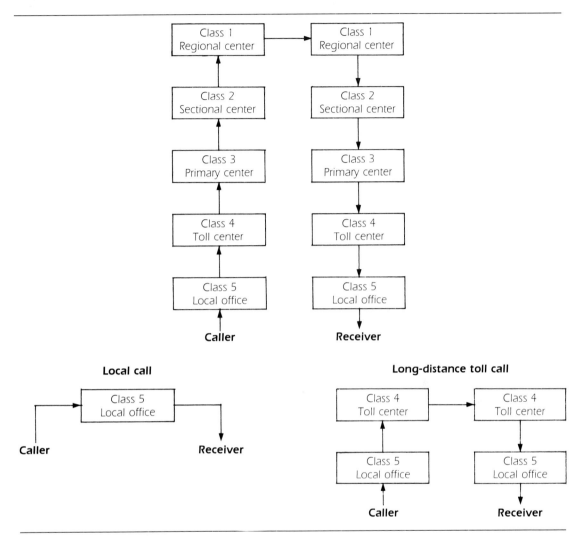

FIGURE 2-2 *Switching office hierarchy.*

a phone is picked up, here's what happens: the scanner detects the voltage change on the line, and the signaling equipment immediately sends out a dial tone. The caller then dials the number and the computer converts these digits into electronic instructions that connect the caller to the appropriate local or trunk lines. At the other end, the phone rings. The time, length, and destination of the call are recorded for billing purposes.

4. *The transmission network.* The long-distance trunk lines (also called **interexchange circuits**) interconnect the central offices and carry voice or data between two local exchanges (to which subscribers are connected by local loops). This traffic may go by twisted-wire pairs, coaxial

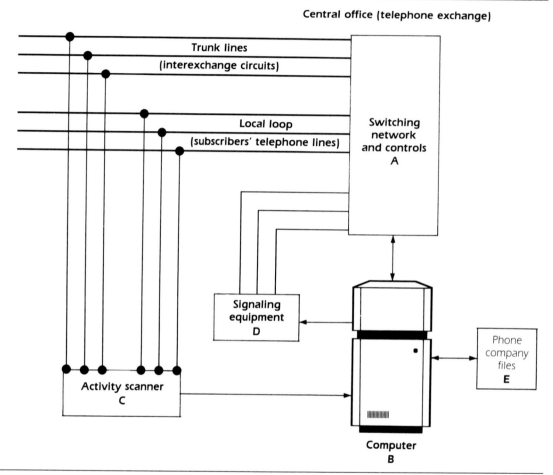

FIGURE 2-3 *Electronic switching system.*

cable, microwave transmission, optical fiber cables, radio waves bounced off satellites, or by a combination of these methods.

Twisted-wire pairs — pairs of copper wires — are the backbone of the communications industry, since they are inexpensive and easy to install. The local loops which connect home phones to the phone company's central offices are normally twisted-wire pairs. Unfortunately, twisted-wire pairs are susceptible to electrical interference and have a relatively slow data transfer rate (i.e., they cannot transmit high frequencies).

Coaxial cable is less subject to interference and can transmit higher frequencies than twisted-wire pairs, but is more expensive to manufacture and install. In cable TV areas, for example, coaxial cables are used to distribute the television signals.

Microwave transmission can take place only within line of sight, so microwave relay towers have to be placed about every 25 miles to offset the earth's curvature. Each relay station amplifies the microwave signals and retransmits them. Microwave can carry thousands of voice channels across lakes, mountains, and deserts without the need to string wires or lay down cables.

ANALOG VS. DIGITAL TRANSMISSION

Information can be transmitted in either analog or digital form:

Analog signals are continuous waves of electrical current. Information is transmitted by varying either the voltage or the frequency of the current. When the voltage of the current is varied, the method of transmission is called **amplitude modulation** (AM). When the frequency of the current is varied, it is called **frequency modulation** (FM) (see Figure 2–4).

FIGURE 2–4 ————————————————————————————

Analog technology.

Unmodulated Carrier Signal

Amplitude

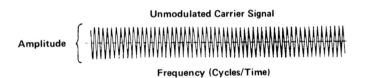

Frequency (Cycles/Time)

Analog Message Signal

Modulated Carrier

Amplitude Modulation

Analog Message Signal

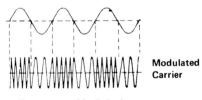

Modulated Carrier

Frequency Modulation

IBM, An Introduction to Local Area Networks (1984), Pub. No. GC20-8203-1, 3–6. Courtesy of International Business Machines Corporation.

Radio broadcasts also use amplitude and frequency modulation, but it is the amplitude and frequency of radio waves, not electrical current, that is varied. Another example of analog technology is cable television networks, which transmit several frequencies over the same cable. The viewer can tune into any of the frequencies by simply changing the channel. Because a broad range of analog frequencies can be transmitted over a single cable, this type of system has come to be called a **broadband** system.

Digital signals do not vary continuously as do analog signals. They have only two conditions: on or off, corresponding to levels of voltage or current. The individual pulses produced by turning the voltage or current on or off are known as **bits,** which stands for *binary digits* (binary is the base two numbering system, which consists of zero and one). Individual bits are sent at a predetermined rate, and a high or low level of voltage or current on the line signifies either zero or one. Patterns of bits are used to represent numerals, characters, system commands, and other information (see Figure 2–5).

In digital technology, several signals can share a single cable by means of **time division multiplexing.** This is becoming an increasingly important technique in communication technology, particularly in computer and satellite networks which must carry a lot of information at one time. Systems that use digital technology are called **baseband** systems, since the digital information is not modulated or changed.

Our telephone system developed using the old-fashioned, less efficient analog transmission, and it remains basically analog because of the huge investment in plant and equipment. But digital technology is rapidly becoming the choice when equipment is replaced or networks are expanded. Before long, high-speed digital transmissions will be the basis of all our communications, and voice, text, data, television, and images will all be coded and sent over digital channels.

TELEPHONE CALLS

Let's look at a typical office situation. Sandi Kirshner, the director of marketing, wants to check on the status of a new product under development. Using the **private branch exchange (PBX),** the company's in-house telephone system, she calls Richard Morel, the head of research and development. Sandi's extension is 123; Richard's extension is 789. Here is the sequence:

1. The handset of extension 123 is picked up and the PBX recognizes that it is off the hook.
2. The PBX generates a dial tone, which stops when the first digit (7) is dialed.

FIGURE 2–5

Digital technology.

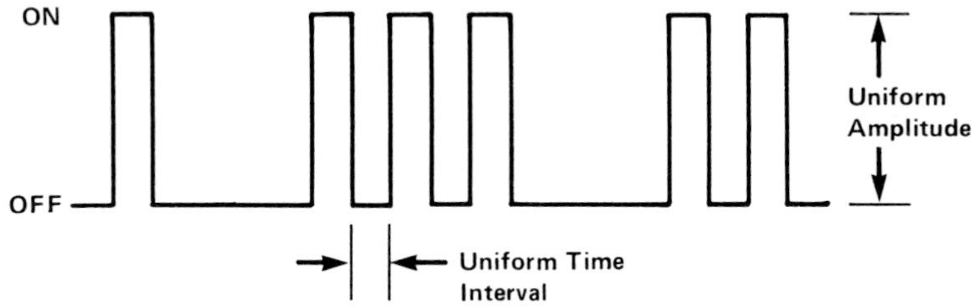

IBM, An Introduction to Local Area Networks (1984), Pub. No. GC20–8203–1, 3–7. Courtesy of International Business Machines Corporation.

3. The PBX stores the digits, and when all three have been dialed, it tests the number 789 for legitimacy. If it is not a "live" number, the PBX generates a special rapid busy signal. If 789 is busy, the PBX sends extension 123 a normal busy signal.

4. Since extension 789 is free, the PBX rings it.

5. When the handset of 789 is picked up, the ringing stops and a connection is established between extensions 123 and 789.

6. The PBX maintains the connection until the handset at either extension is hung up.

Now that Sandi has learned the exact delivery date of the new product, she needs to inform a potential customer of this date. A call from her extension to the prospective buyer's outside number, 999–2931, follows these steps:

1. The PBX recognizes that extension 123 has been picked up and generates a dial tone.

2. An outside access code number (e.g., 9) is dialed.

3. The PBX finds a free telephone line to the local phone company, and sends out an *off-hook* signal.

4. The local phone company office returns a dial tone until the first digit is dialed.

5. The PBX stores the dialed number (999–2931) and then sends it down the line to the phone company.

6. The connection is broken when extension 123 is hung up or the PBX receives an *on-hook* signal from the phone company.

The customer is out of the office, but soon calls Sandi back:

1. The local phone company office sends the PBX a signal that a call is coming in on a telephone line.

2. The PBX sends a "send the number" signal down the line.

3. The PBX receives the number 123, checks it for legitimacy, and checks to see if the extension is busy.

4. Since extension 123 is free, the PBX rings it.

5. The connection is broken when extension 123 is hung up or the PBX receives an on-hook signal.

ELEMENTS OF COMMUNICATION

A telephone system illustrates the basic elements of communication:

1. **Source.** The source is the individual who needs to send a message to someone else, whether by letter, telephone, or any other method.

2. **Transmitter.** The transmitter changes the message into the form required for delivery. For instance, the mouthpiece of a telephone is a transmitter, converting sound into electronic signals.

3. **Channel.** The channel is how the message moves from source to destination. In a telephone call, the channel includes all the wires, phone lines, and switches between the two handsets.

4. **Receiver.** The receiver changes the message into a form understandable by the recipient. A telephone handset, for example, converts electronic signals back into sound.

5. **Destination.** The destination is the individual for whom the message is intended. Communication takes place only when the destination receives the message sent by the source.

In a data communications system, these elements take the form of the components illustrated in Figure 2–6. Before we take a look at how these components work, let's briefly discuss the codes used to transmit information.

CODES

Every driver knows that a red traffic signal means "stop," that green means "go," and that yellow signifies "caution." This is a *code:* a representation of information in a form that is different from its original form, but one that can be understood by both the sender and receiver.

In business telecommunications, we usually need to transmit a stream of numbers, letters, and special characters, so the information has to be converted from a form people understand to something the network can handle, and then has to be converted back when it reaches destination.

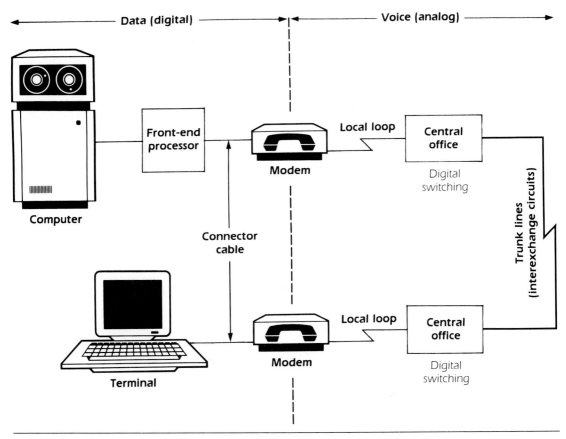

FIGURE 2–6 Data communications system components.

Electronic equipment, however, has only two codes to work with: on or off, current flowing one way or the other, elements that are magnetized or not. We can use the binary system to describe these two states: for example, the presence of an electric current can be signified by the number one, and the absence of a current can be represented by zero. We can thereby encode a message into a string of ones and zeros that can be transmitted along a channel for decoding by the receiver. All we need is a code to represent the alphabet, numbers, and special characters as ones and zeros.

We use the decimal system, with its ten digits zero through nine, because we have ten fingers and ten toes. But even though the binary system has only two digits, zero and one, it can also do everything we need. In data communications, a binary digit or bit is a zero or a one—the smallest unit of information. To represent a single character, such as the number three or the letter "T," you need eight binary digits or bits—in computer terminology, this is called one *byte*. So, each time a computer transmits a number or a character, it is really dealing with eight separate "bits" of information. Figures 2–7, 2–8, 2–9, and 2–10 illustrate how numbers and letters can be encoded as bits.

ASCII

The American National Standard Code for Information Interchange (ASCII or ANSCII) consists of seven information bits, plus an eighth bit for checking **parity.** Parity simply means whether the sum of all the bits is odd *(odd parity)* or even *(even parity)*, and helps the receiver to check for transmission errors (see the bottom of Figure 2–7). ASCII is widely used for data transmission, and although there are a number of standardized versions, they are all basically as illustrated in Figure 2–7. International characters can be easily incorporated: For instance, in the United Kingdom, the pound sign (£) is used instead of the dollar sign ($). The precise definitions of all the characters are determined by the various national standards organizations (such as the National Bureau of Standards).

In addition to the letters, numbers, and punctuation marks, there are 32 control characters in the ASCII character set. These can be classified into four groups (see Figure 2–8):

1. **Transmission controls** are used to control the flow of data along the channel.
2. **Information separators** are used to define the various parts of a message.
3. **Device controls** are used to control particular terminals and to determine whether a message was correctly received.
4. **Format controls** are used to control the physical layout of information on the printed page or the video display screen.

Every user does not need all these control codes and special characters. Some users redefine various control characters for their own applications, but when two such users of a modified ASCII code try to communicate, there may be problems. For example, the eighth ASCII bit is normally the parity bit, but some terminals use it to define special functions such as accessing the disk drive or setting the printer format. Such a terminal, therefore, may not be able to communicate properly over a telecommunications network. In general, it's a good idea to leave the ASCII code the way it is shown in Figure 2–7.

EBCDIC

The Extended Binary Coded Decimal Interchange Code (EBCDIC) is another widely used eight-bit code (see Figure 2–9). It is the internal machine code used in many computers, and therefore is often the data transmission code

FIGURE 2-7

ASCII code set.

Bits 4 3 2 1	7 6 5 → 0 0 0	0 0 1	0 1 0	0 1 1	1 0 0	1 0 1	1 1 0	1 1 1
0 0 0 0	NUL	DLE	SP	0	@	P	\	p
0 0 0 1	SOH	DC1	!	1	A	Q	a	q
0 0 1 0	STX	DC2	''	2	B	R	b	r
0 0 1 1	ETX	DC3	#	3	C	S	c	s
0 1 0 0	EOT	DC4	$	4	D	T	d	t
0 1 0 1	ENQ	NAK	%	5	E	U	e	u
0 1 1 0	ACK	SYN	&	6	F	V	f	v
0 1 1 1	BEL	ETB	'	7	G	W	g	w
1 0 0 0	BS	CAN	(	8	H	X	h	x
1 0 0 1	HT	EM	)	9	I	Y	i	y
1 0 1 0	LF	SUB	*	:	J	Z	j	z
1 0 1 1	VT	ESC	+	;	K	[	k	{
1 1 0 0	FF	FS	,	<	L	\	l	:
1 1 0 1	CR	GS	-	=	M	]	m	}
1 1 1 0	SO	RS	.	>	N	∧	n	~
1 1 1 1	SI	US	/	?	O	—	o	DEL

Example:

Bits: P*7 6 5 4 3 2 1

1 1 0 0 0 0 0 1 = letter "A" (Odd Parity)

0 0 1 1 1 0 0 0 = number "B" (Odd Parity)

P* = Parity Bit

Ken Sherman, **Data Communications: A User's Guide**, 2nd ed. © 1985, p. 95 (A Reston Publication). Reprinted by permission of Prentice-Hall, Inc., Englewood Cliffs, New Jersey.

for these computers as well. The above cautions also apply: even though the computers are using EBCDIC, they may not be able to talk to each other if some of the character assignments have been modified.

The bit positions are different for ASCII and EBCDIC, as illustrated in Figure 2–10. Therefore, if you are converting an accounts receivable file from EBCDIC to ASCII, and if bit number two in EBCDIC shows whether or not the credit limit has been exceeded, this bit will become bit number six when the file is converted into ASCII.

In summary, codes such as ASCII and EBCDIC are used to translate messages and other data into bits, which can then be sent over data communication systems.

FIGURE 2-8 ———

ASCII control
characters.

Transmission Controls

NUL (Null)	The all zeros character, used for time or media fill.
SYN (Synchronous Idle)	Used for character synchronization in synchronous transmissions.
DEL (Delete)	Used to erase in paper tape punching.

Information Separators

SOH (Start of Header)	Used at the beginning of routing information.
STX (Start of Text)	Used at the end of header or start of text.
ETX (End of Text)	Used at end of text or start of trailer.
EOT (End of Transmission)	Used at end of transmission (i.e., end of call).
SO (Shift Out)	Code characters which follow are not in the code set of the standard code in use. (Predefined as to which code you shift to.)
SI (Shift In)	Code characters which follow are in the code set of the standard code in use.
DLE (Data Link Escape)	Used to change the meaning to a limited number of contiguously following characters.
ETB (End of Transmission Block)	Used to indicate end of a block of data.
CAN (Cancel)	Disregard the data sent with.
EM (End of Medium)	End of wanted information recorded on medium.
SS (Start of Special Sequence)	
ESC (Escape)	Used to extend code.
FS, GS, RS, US (File, Group, Record, Unit Separators)	

Device Controls

ENQ (Enquire)	Used as a request for response: "Who are you?"
ACK (Acknowledge)	Used as an affirmative response to a sender.
BEL (Bell)	Used to call for human attention.
DC1, DC2, DC3, DC4 (Device Controls)	Characters for the control of auxiliary devices (i.e., start, pause, stop).
NAK (Negative Acknowledge)	Used as a negative response to a sender.

Format Controls

BS (Backspace)	Moves a printing device back one space on the same line.
HT (Horizontal Tab)	Moves a printing device to the next predetermined position along a line.
LF (Line Feed)	Moves a printing device to the next printing line.
VT (Vertical Tab)	Moves a printing device to the next predetermined printing line.
FF (Form Feed)	Moves the paper to the next page.
CR (Carriage Return)	Moves the printing device to the left margin.
SP (Space)	A format effector, used to separate words.

Adapted from Ken Sherman, **Data Communications: A User's Guide,** 2nd ed. © 1985, pp. 96–97 (A Reston Publication). Reprinted by permission of Prentice-Hall, Inc., Englewood Cliffs, New Jersey.

Bits 4567 ↓	Hex1	000				01				10				11				Bits 0,1
		00	01	10	11	00	01	10	11	00	01	10	11	00	01	10	11	2,3
		0	1	2	3	4	5	6	7	8	9	A	B	C	D	E	F	Hex 0
0000 0		NUL	DLE			SP	&	-									0	
0001 1		SOH	SBA				/			a	j			A	J		1	
0010 2		STX	EUA		SYN					b	k	s		B	K	S	2	
0011 3		ETX	IC							c	l	t		C	L	T	3	
0100 4										d	m	u		D	M	U	4	
0101 5		PT	NL							e	n	v		E	N	V	5	
0110 6				ETB						f	o	w		F	O	W	6	
0111 7				ESC	EOT					g	p	x		G	P	X	7	
1000 8										h	q	y		H	Q	Y	8	
1001 9			EM							i	r	z		I	R	Z	9	
1010 A						¢	!	|	:									
1011 B						.	$	,	#									
1100 C			DUP		RA	<	.	%	@									
1101 D			SF	ENQ	NAK	(	)	—	'									
1110 E			FM			+	;	>	=									
1111 F			ITB		SUB	|	—	?	''									

EBDCIC Code as Implemented for the IBM 3270 Information Display System
Extended Binary Coded Decimal Interchange Code (IBM's EDCDIC).
a. An 8-bit code yielding 256 possible combinations or character assignments.
b. A representative subset, that of the IBM 3270 product family.
  Absence of certain functions not usable by 3270 products (e.g., paper feed, vertical tab, back space) which would show up in EBCDIC code charts for other products that might make use of them.

FIGURE 2–9   EBCDIC code set.

Jerry FitzGerald, **Business Data Communications**, p. 60. Copyright © 1984 John Wiley & Sons. Reprinted by permission of John Wiley & Sons.

──────── TERMINALS ────────────────

One of these terminal types can probably be found at the end of any data communication system:

1. **Teleprinters,** or Teletypes, have keyboards and produce paper print-outs. They have no programming capability.

FIGURE 2-10   *ASCII and EBCDIC bit positions.*

2. **CRTs.** Cathode ray tubes (CRTs), or video display terminals, have television screens and keyboards. CRTs are very common.

3. **RJE terminals** are used for remote job entry (RJE). Data is entered at a remote location, and then sent over telephone lines or other channels to a central processing unit, a centralized mainframe computer. The results are then transmitted back to the RJE terminal for printing.

4. **Intelligent terminals** are really small computers: they have microprocessors which can be programmed and which can execute stored programs.

Whatever its type, a terminal uses digital transmission of electrical pulses: bits are represented by plus and minus voltages, such as + 15 volts for one and −15 volts for zero. Assuming the ASCII code is used, when you type the number eight on your keyboard the terminal outputs eight bits: 00111000 (see the bottom of Figure 2–7). We call this kind of terminal output **baseband** signals, since the original digital signal is not modified until a modem converts it into an analog signal (see below).

The encoded characters — the eight-bit groups, or bytes — are transmitted **serially** (one at a time) along the channel. The receiver (the central computer in Figure 2–6) has to determine which group of bits belongs to what character, and it does this by counting off the bits eight at a time and assembling each character. In contrast, **parallel** transmissions send all eight ASCII bits simultaneously over eight channels, and all eight bits are received together as a single character. This results in a higher data transfer rate, but it also requires eight separate channels, so parallel transmission is normally used only within computer systems (for instance, between the disk drives and the central processing unit).

There are two ways to determine which is the first bit of an incoming character: asynchronous and synchronous transmission.

■ **Asynchronous** transmission is used when the characters are sent one at a time, usually without a fixed interval between characters. Start and stop bits are placed at each end of each individual eight-bit character to separate the characters and to synchronize transmission. A character is transmitted whenever the operator presses a key, and the receiver is activated when it detects a start bit and operates for as many bits as there are in a character. Then it idles until the next character is transmitted. Asynchronous transmission is used for interactive time-sharing, since it allows many terminals fast access to a central computer.

■ **Synchronous** transmission is used for high-speed transmission of a block of characters. Rather than using start and stop bits, synchronization is achieved by sending **synchronization characters** (or SYN characters) ahead of each block of message bits. The block of data may consist of several thousand individual bits, all uniformly spaced, and might represent a whole line or screen on a terminal. Synchronous transmission is used only by *buffered terminals*—terminals with the ability to store data in a **buffer,** so it can be accumulated and sent as a block rather than one character at a time. The SYN control characters precede each block and establish a timing pattern between the sender and receiver.

———— CONNECTOR CABLES ————————————————

Manufacturers of computers, terminals, and modems all have to make sure their products interface properly with data communications equipment and services. Fortunately, the Electronic Industries Association in conjunction with the telephone company and modem and computer manufacturers has developed a serial binary interchange standard called *RS 232C* ("C" is the latest revision). It is the standard for which all modem/terminal and modem/computer interfaces are designed, and it defines the physical design of the plugs and outlets that connect user devices to modems.

The RS 232C interface is a connector with 25 pins. (A newer interface called the RS 440 works on the same principle but has 37 pins with an optional nine-pin connector.) Each pin is responsible for carrying a specific type of signal from the modem to the terminal or computer, or from the terminal to the modem, and each signal has a different voltage level. Thus, the interface not only allows mechanical compatibility, but electrical compatibility as well.

The terminal or computer is attached to the modem by a connector cable (see Figure 2–6), a 25-wire cable up to 50 feet long. The plug has 25 pins, as illustrated in Figure 2–11. For example, the second pin carries the digital data from the terminal or computer to the modem, the third pin carries the digital data from the modem to the terminal, the fourth pin signals the modem that the terminal has data to transmit, and the fifth pin signals the terminal that the modem is ready to receive.

A "handshaking" sequence, which occurs automatically between the modems, establishes that a circuit is available and open for transmissions (we shall discuss this process later).

FIGURE 2–11———
RS 232C interface.

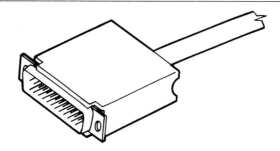

Pin	Function	Pin	Function	Pin	Function
1	Frame Ground	10	Negative dc Test Voltage	19	Sec. Request to Send
2	Transmitted Data	11	Unassigned	20	Data Terminal Ready
3	Received Data	12	Sec. Data Carrier Detect	21	Signal Quality Detect
4	Request to Send	13	Sec. Clear to Send	22	Ring Indicator
5	Clear to Send	14	Sec. Transmitted Data	23	Data Rate Select
6	Data Set Ready	15	Transmitter Clock	24	Ext. Transmitter Clock
7	Signal Ground	16	Sec. Received Data	25	Busy
8	Data Carrier Detect	17	Receiver Clock		
9	Positive dc Test Voltage	18	Receiver Dibit Clock		

The terminal connection to the modem is defined by the Electronic Industries Association (EIA) specification RS 232C. RS 232C specifies the use of a 25-pin connector and the pin on which each signal is placed.

——

Jerry FitzGerald, **Business Data Communictions,** Copyright © 1984 by John Wiley & Sons. Reprinted by permission of John Wiley & Sons.

——————— ## MODEMS———

To use analog telephone lines for digital data transmission, the digital signals must be converted into an analog form that can be sent along the phone circuits. At the other end, they can be converted back into the digital signals used by terminals and computers. That is, digital pulses must be **modulated** into an analog signal.

A **modem** (short for modulator/demodulator) does exactly that. Let's assume your terminal is in one building and a mainframe computer is in another building a few miles away. When you need to use the computer, you dial a local phone number and you hear a continuous, high-pitched noise over the telephone handset—the **carrier** signal. Then, you place the handset on the **acoustic coupler** of your modem, or you flick a switch on

This section on modems was adapted from Wang Laboratories, *A Primer on Computer Networks* (1982), No. 777–4004, 21–23.

your direct-connect modem. Your modem is now connected to a **port** on the computer: an access point for data entry or exit through a single data channel.

Remember that our telephone system was designed to carry continuously variable voice signals and not on-or-off digital pulses. These analog voice signals travel in alternating wave patterns, and one full "wave" from zero to plus through zero to minus and back to zero is called a *cycle* (see Figure 2–12). A cycle is also known as a *sine wave*, and sine waves are measured by their amplitude (strength), frequency, and wavelength.

Amplitude increases or decreases according to the voltage that is applied to the telephone wire. The distance between the wave crests is called the *wavelength*. The number of times a cycle repeats itself in one second is the *frequency* of the wave, measured in cycles per second (cps) or **hertz** (Hz). Thus, a wave with 1,000 cycles per second has a wavelength of 1,000 hertz or one kilohertz.*

As noted earlier, the continuous analog signal that modems use to communicate over the telephone lines (and which we hear as a high-pitched whine) is known as a carrier signal or carrier wave. To transmit your digital data over a telephone line, the carrier signal of your modem must be modified or modulated by the data. The modem takes the digital pulses and *impresses* them onto the carrier wave, creating an analog signal (see Figure 2–13).

At the other end of the line, another modem separates or demodulates the digital pulses from the carrier wave and passes them on to the receiving device.

There are several methods by which individual digital pulses can alter the physical form of a carrier wave. The two most common methods, as we mentioned earlier, are amplitude modulation (AM) and frequency modulation (FM).

FIGURE 2–12
Cycle.

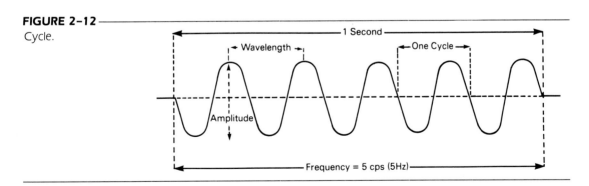

*1 Hz (hertz) = 1 cycle per second
1 KHz (kilohertz) = 1,000 cycles per second
1 MHz (megahertz) = 1 million cycles per second
1 GHz (gigahertz) = 1 billion cycles per second.

FIGURE 2-13

Analog signal.

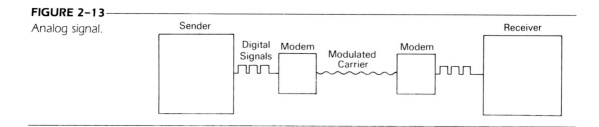

Amplitude Modulation

In this method, the amplitude or voltage of the carrier signal is modulated by the value of the bits being transmitted. A binary zero results in a small wave (measured from top to bottom), while a binary one produces a bigger wave (see Figure 2–14). This is also known as *amplitude shift keying* (ASK).

Frequency Modulation

In frequency modulation, the frequency of the carrier signal is modulated by the value of the bits being transmitted. For example, a zero might be represented by a frequency of 1,000 Hz (the longer wave cycle in Figure 2–15), while a one would be represented by a frequency of 1,500 Hz (the shorter cycle). This technique is known as *frequency shift keying* (FSK), and is more commonly used in digital-to-analog transmission systems.

Bits Per Second and Baud

Since the smallest unit of digital information is the bit, we refer to the capacity of a channel to carry data in terms of *bits per second* (bps). Unfortunately, bits per second is sometimes confused with the **baud** rate, which normally means the number of individual signal changes per second. Bps

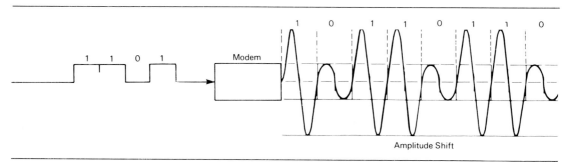

FIGURE 2-14 *Amplitude shift keying.*

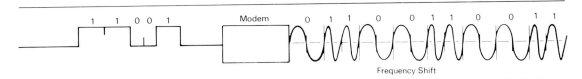

FIGURE 2-15 Frequency shift keying.

and baud rate are identical only when the modem transmits one bit per cycle of the carrier wave. If, however, four bits are transmitted per cycle, you might have a 9600 bps modem transmitting at a baud rate of 2400. Bps is what we will use in this book, and the term baud is discouraged. We are interested in bps because bits make up characters and characters make up the words used in business telecommunications.

——— THE LOCAL LOOP ———

We have seen that the local loop connects telephone subscribers to the local telephone company switching facility. The telephone jack on your office wall which you plug your phone into is one end of the local loop; the other end is the phone company central office (see below). Private **leased lines,** unlike normal telephone lines, are reserved for the sole use of a single customer and avoid the switching facility entirely. Leased lines may be less expensive than normal switched lines, and can move a greater volume of data faster and more accurately.

Before the data reaches the modem, signal quality and errors are the user's responsibility. But once the signal is past the modem and into the local loop, the common carrier—the phone company—is responsible for the circuit quality and for any errors. The long-distance carrier is responsible for signal quality after it is routed by the central office into the interexchange circuits (see Figure 2-6).

The signals transmitted over the local loop and interexchange circuits may be either half-duplex or full-duplex:

- **Half-duplex** channels can transmit in either direction, but only in one direction at a time. A one-lane bridge is half-duplex, and so are walkie-talkies that can either send or receive, but not both simultaneously.

- **Full-duplex** channels can transmit in both directions simultaneously. Most roads, for example, are full-duplex, as are most telephone conversations. Providing full-duplex capability may improve the efficiency of a system (although probably not when both people talk at the same time during a phone call). Full-duplex and half-duplex will be discussed later when we consider protocols and line disciplines.

THE CENTRAL OFFICE

The telephone company central office is where circuit switching occurs for normal station-to-station calls. Leased lines, as we mentioned earlier, are private phone lines that are wired around the switching gear, which means less noise and fewer errors because switching is eliminated and the lines can be conditioned for better transmission. Leased lines are also called *dedicated lines* or *tie lines*.

The central office is where the local loops end and where the long-haul circuits begin. Since these long-distance lines connect the central exchanges, they are termed **interexchange circuits** (IXCs). The IXCs may use wire pairs, coaxial cables, microwave transmitters, satellite networks, or whatever technology is in place. Interexchange circuits are also called *public-switched*, *direct-distance dialing* (DDD), and *message toll service* (MTS) lines.

THE FRONT-END PROCESSOR

At the other end of the interexchange circuit, the signal passes through another central office, local loop, modem, and connector cable before it reaches the **front-end processor** (FEP). The FEP relieves the central computer (see below) of a number of routine functions required to control the telecommunications network, such as line control, message handling, code conversion, and error control. The front-end processor also strips incoming messages of their communication control characters, and sends them "naked" to the central computer for processing.

THE CENTRAL COMPUTER

The central computer has many names: **mainframe, central processing unit** (CPU), and host computer (the latter because large data bases are often stored in its peripherals). When the signal leaves the front-end processor, it is processed by the computer, checked for any security restrictions, and if it checks out, the instructions are implemented. The incoming signal may, for example, update a data base field, or retrieve stored information.

To summarize: our telephone system was made for talking. It is not ideal for the transmission of data because it was designed and engineered for continuously varying analog signals. As the need to transmit digital data has increased, a technique has been developed to send digital data over phone lines using modems. At each end, before and after the modems, the signal is digital (baseband). Between the modems, the signal is analog (broadband). We are thus able to transmit digital data with the installed analog (voice) lines. New equipment tends to be all-digital, eliminating the need for special equipment like modems; but the conversion is necessarily slow, so the techniques discussed in this chapter will be used for quite a while.

The article reprinted at the end of this chapter, "Micros Slash Firm's Telecom Expenditures," discusses how the Federal Mogul Corp., a multinational manufacturing firm, came to use G.E. Information Services' international time-sharing network, and illustrates the use of a network other than the telephone company's.

THE BRISFIELD COMPANY

After reviewing the Brisfield Company case at the end of Chapter 1:

1. Describe in your own words the remote job entry (RJE) applications discussed under "Remote Computing" in the Brisfield Company case.
2. Describe the functions of the modems used by the Brisfield Company in its early days of remote computing.

SUMMARY

1. Our analog (voice) telephone network has four basic parts: telephones, local loops, switching centers, and the transmission network.
2. The existing analog equipment is being replaced and expanded with digital equipment.
3. We are moving toward all-digital networks which can transmit voice, text, data, and images fast, accurately, and inexpensively.
4. We use the telephone system to transmit either voice or data in roughly the same way, except that digital data must first be conversion into analog form.
5. Electronic equipment can be characterized as having two states, on and off, so it uses the binary numbering system: zero and one.
6. Two widely used binary representations of our numbers and letters are the ASCII and EBCDID codes.
7. Terminals transmit asynchronously (one character at a time) or synchronously (a block of characters at a time).
8. The interface between the terminal or computer and the modem is the RS 232C, a standard serial interface.
9. Modems convert digital data into an analog signal that can be sent over the telephone lines, and convert the analog signal back into digital data at the other end.

10. Modems use amplitude and frequency modulation to modulate the carrier wave and transmit the digital data.

11. Bits per second and baud both measure the rate of data transmission, but we will use only bps in this book.

12. Local loops connect telephone subscribers to the switching gear at the telephone company central office.

13. Interexchange circuits connect the central offices.

14. Front-end processors relieve central computers from having to control the telecommunications network.

REVIEW QUESTIONS

1. Discuss the basic parts of our telephone system.

2. What part does software play in the switching function?

3. Explain why the analog system is being replaced or expanded with digital technology.

4. Describe what happens when you call extension 789 from extension 123 on the same computer branch exchange.

5. Describe what happens when you call an outside number from extension 123.

6. How is sending data over the telephone lines different from sending voice?

7. Why do computers use binary codes?

8. What are ASCII and EBCDIC? Give examples.

9. What are control characters?

10. What is asynchronous transmission? Name a typical use.

11. What is synchronous transmission? Name a typical use.

12. What is a RS 232C interface? Why is it important?

13. What does a modem do? How?

14. What is a hertz? Kilohertz? Megahertz? Gigahertz?

15. What is the difference between bits per second (bps) and baud? Why do we prefer bps?

16. What is the local loop? What is a central office?

17. Define half-duplex and full-duplex.

18. What is the function of a front-end processor?

19. Why is a mainframe sometimes called a host computer?

20. How did the Federal Mogul Corp. reduce its communication costs with the G.E. network?

MICROS SLASH FIRM'S TELECOM EXPENDITURES

Switch from Dumb Terminals Pays Off for Multinational

SOUTHFIELD, Mich.—A multinational manufacturing firm based here cut its monthly telecommunications costs for international financial reporting from $25,000 to $2,000 when it began using microcomputers rather than dumb terminals to access time-sharing services.

Federal Mogul Corp. makes ball bearings, sealing devices, and other parts used in the trucking, construction, and aerospace industries and general industrial applications. The firm has manufacturing plants in 10 foreign countries and maintains inventories in 20 warehouses overseas. Its net sales in 1984 were $912 million; international sales accounted for $243 million of the total.

Until November 1983, employees at Federal Mogul's foreign offices keyed financial data about their operations into dumb terminals that were linked on-line to mainframes on General Electric Information Services Co.'s (Geisco) international time-sharing network.

Workers in the firm's international finance department here accessed and processed the information by dialing dumb terminals into a Geisco mainframe in Rockville, Md.

"Using dumb terminals meant that the mainframe did the actual number crunching," according to Wilhelm A. Schmelzer, Federal Mogul's director of international finance. "With each of roughly 30 international locations inputting various types of data . . . this led to intolerable costs." The firm's headquarters was spending about $25,000 per month for line charges, CPU time, and data storage.

The foreign offices were paying between $800 and $3,000 per month in line and connect charges. Employees in each office had to go on-line in order to key data into the system. "That is where the real cost came in," Schmelzer said. The costs varied according to how much time individual employees took to type in data, check their work, and make corrections.

After a year's experience with the time-sharing setup, corporate management deemed the charges excessive and instructed the international finance department to cut costs. Schmelzer sent representatives to the foreign offices to check their procedures. He then analyzed the time-sharing bills to see where the firm could save money.

"WE HAD TO TAKE OTHER MEASURES"

In a series of stopgap measures, his group pulled some foreign offices off the network, eliminated some types of financial reporting, and cut back on the frequency of some transactions. The investigations and service cutbacks did not yield sufficient savings. "That's when we finally found that . . . we had to take other measures," Schmelzer said.

In September 1983, the department set up two trial sites for microcomputer-based time-sharing. The trial sites, Federal Mogul offices in Puerto Rico and South Africa, received IBM Personal Computers fitted with modems and financial software.

Employees in the two offices entered financial data on the Personal Computers and used the software, IMRS Inc.'s Micro Control, to convert the figures from local currencies to U.S. dollars. After they entered, checked, and corrected their work, the employees used the software's communications ability to dial in to the Geisco network and transmit their data.

The tests paid off. Because the offices entered and converted their data on the local processors and went on-line only to transmit the information, their combined line charges dropped from about $2,300 to about $600. The managers of the local offices were pleased with the results, as were members of the international finance department. According to Schmelzer, "We jumped through the roof here."

Schmelzer's group recommended to corporate management that Federal Mogul adopt the micro-based setup for all its international accounting transactions. In November 1983, the firm began to install IBM Personal Computers and compatible machines, each running the Micro Control software, in other foreign offices and at its headquarters. Twenty-three Federal Mogul facilities in 18 countries currently use the system; one more is scheduled to come up soon.

The firm's headquarters brought its monthly time-sharing costs down to about $2,000 a month, about $1,200 of which constitutes charges for data storage. The foreign offices' average time-sharing costs have dropped to between $25 and $100 per month.

With the system, the foreign offices key in financial data on their personal computers and then go on-line to transmit the information to a mainframe on the Geisco network. Employees in the international finance department at the firm's headquarters then dial in to the time-sharing network to call up the information.

They download the data to an IBM Personal Computer XT and use the Micro Control software to compile the information into financial statements and management reports. "We still use the communications link we used prior to installing Micro Control," Schmelzer said. "It's just that every computation we need is done at the office concerned, and the link through the network's time-share is purely to transfer data."

The cost savings from local processing have eased relations between the international financial department and the foreign office managers, who have to foot the bill for transmitting their accounting data.

The managers had complained about the cost of carrying out reporting procedures with the terminals. Although they complied with corporate guidelines, they did so grudgingly, according to Schmelzer. "You can imagine the support we have now from the managers," he said.

In the foreign offices, employees use their personal computers for tasks other than transmitting financial information. Some small offices have set up inventory and receivables systems, and all the offices use Lotus Development Corp.'s 1-2-3 spreadsheet package, according to Schmelzer's assistant, Mark Gergel, Federal Mogul's international accounting manager. "We've come a long, long way in one year in automating those locations," Gergel said.

By bringing microcomputers to the foreign offices, the international reporting system provided the basis for these other uses of automation, he said. "If it weren't for the . . . reporting system, we'd be light-years behind."

The system's software has given the international finance department a variety of reporting capabilities. Schmelzer's group uses the software to compile 80 separate ledgers each month, including profit-and-loss statements and balance sheets for each foreign location.

QUICK CLOSING
With the international reporting system, the department can close its books for any given month within 11 working days of the month's end, Schmelzer said.

At the end of each month, the department transfers compiled financial data to Federal Mogul's corporate general ledger system, which runs on two IBM mainframes: a 3033 and a 3084 Model QX. A terminal emulation board on the IBM Personal Computer XT allows for the transfer.

Gergel said one of the software's greatest benefits is that it allows the department to format reports in-house. When Federal Mogul used Geisco to process its financial data, it had to request format changes and wait for the service bureau to make them. Most requests required a turnaround time of between 10 days and 3 weeks.

An in-house programmer can now develop reports on the microcomputer in as little as 15 minutes, he said. With in-house development capability, Gergel said, "We're no longer reliant on anybody."

3 DIGITAL TELECOMMUNICATIONS

BACKGROUND
TERMINALS
SWITCHES
TRANSMISSION

INTEGRATED SERVICES
DIGITAL NETWORKS
OPTIMIZATION

AFTER READING THIS CHAPTER, YOU WILL UNDERSTAND THAT:

- Our existing telephone network uses predominantly analog technology, but is shifting to digital technology as it is replaced and expanded.

- The telecommunications network has three major parts: terminals, switches, and transmission facilities. Each part is being replaced with digital technology at a different rate.

- The office private branch exchange is the hub of the modern automated office.

- The proportion of digital data on the telephone network is increasing.

- Integrated services digital networks will use digital technology to offer an array of services on a global scale.

As noted in Chapter 2, the United States enjoys excellent telephone service. Since the telephone network is both available and extensive, it has been called upon to transmit an increasing volume of digital data generated by a growing number of computers. As we saw, however, computer and telephone technology are not ideal bedfellows. Chapter 3 discusses this compatibility problem as well as the other fundamentals of digital communications.

All computer-related devices, including mainframes, terminals, minicomputers, and microcomputers, are digital. Back in the late 1950s, when large organizations had to transfer data between locations, they shipped reels of magnetic tape encoded with the digital data. A slightly more efficient system was batch transmission terminals, which often used leased lines at night to transmit data from one reel of magnetic tape to another (via a pair of modems), thereby keeping the lines free for voice communications during the day. The copy of the tape could then be processed by the central computer.

Next, companies started transmitting data directly from terminal to computer, rather than copying the data onto magnetic tape first. The technology that made this possible was multiprogram operating systems, which allowed the mainframes to use a segment of their main memory to control these transmissions while they ran other applications in other memory partitions. The transmitted jobs were stored in order (or *queued*) on a disk for later processing, after which the results were transmitted back to the terminal. The IBM 2780 terminal, which used the Bisync protocol, was an early 1960s example of a remote batch-processing system (see Chapter 6).

Time-sharing was developed in the mid-1960s, and soon expanded to include interactive application programs and data bases (e.g., programs to update accounts receivable and inventory whenever a sale was made). The early, slow teletype terminals were replaced by faster video display terminals.

Again, much of the digital data generated by these interactive computer applications was transmitted over the voice telephone network. Personal computers, electronic mail, and other new technologies all increased the proportion of digital data on the telephone network. This is one reason for the move toward digital forms of transmission. Some others[1] are:

- Digital signals tend to have fewer transmission errors than analog signals, since only the absence or presence of a pulse (a binary zero or one) needs to be detected.
- Very large-scale integration (VLSI) circuitry now makes digital technology economical.
- In a digital network, all data is reduced to a stream of zeros and ones. You can therefore use a single network for voice, data, text, and images.
- Switching and control signals are more efficient and reliable with digital technology.

—————— ## BACKGROUND ————————————————————————————

Historically, analog technology was predominant in telecommunications because it was the least expensive way to transmit voice. Voice was transmitted in analog form to the private branch exchange in any large office and then switched and transmitted to the telephone company's local switching facility. There it was switched again and transmitted as an analog signal to its destination.

That was before 1960 and the development of semiconductors, solid-state electronics, and computers. In those early days, the transmission of data was a minor concern, and telephones were required only to transmit voice. Accordingly, most business telephone equipment is analog, and its cost must be recovered slowly through regular use before it can be written off and replaced. This is why our telecommunications network lags somewhat behind our current technology.

Since the 1970s, however, voice, data, text, and image have all been converging. Today, all transmissions can be sent in digital form (in contrast with analog voice-only signals), and then switched and retransmitted as streams of binary digits. This digital technology tends to cost less to install and maintain than its analog counterpart, and is more reliable. For example, with the advent of very large-scale integration circuitry, the cost of the electronics is but a small part of the local cost of the system.

Accordingly, as the telephone companies expand and replace existing facilities, they are moving toward an all-digital network in which analog voice transmissions will take place only in the telephones at each end. Everything in between will be digital, as illustrated in Figure 3–1.

Our telecommunications network is composed of three major parts: terminals, switches, and transmission facilities (see Figure 3–2). Each part of the network is being replaced with digital technology at a different rate, and the rest of this chapter discusses some important aspects of *digitization* — the conversion of terminals, switches, and transmission networks to digital technology.

—————— ## TERMINALS ————————————————————————————

In this context, **terminals** means telephone handsets and computer-related equipment, including personal computers. Telephones are now almost all analog, and data-processing terminals are all digital.

Your vocal cords generate sound waves when you speak, sound waves that vary in frequency from a few hundred cycles per second to several thousand cycles. The microphone in your telephone handset converts those sound waves into electrical voltages, which vary continuously with your voice frequencies. The voice digitization[2] is done by *coder/decoder* circuitry (also called *codec chips*) within the handset, which transform analog

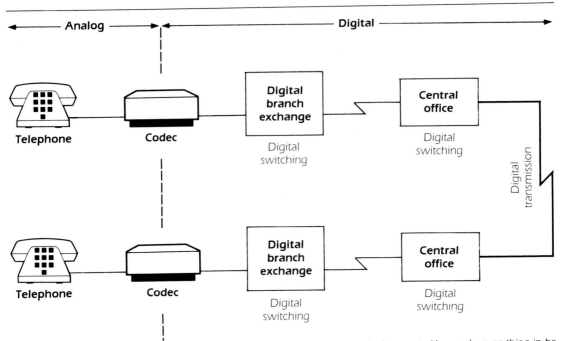

Voice is transmitted in analog form only inside the telephones at either end—everything in between is digital. The speaker's voice is converted from an analog signal to a digital signal by a digital telephone, and the digital signal is transmitted to the office digital branch exchange (DBX). The DBX switches the signal digitally, and transmits it to the telephone company's central office. There the signal is transmitted digitally through the telephone network until it reaches the recipient's DBX. The DBX once again switches the signal digitally and transmits it to the receiving telephone, where it is converted back into analog form for the listener.

FIGURE 3-1 All digital telephone network.

input into a digital bit stream (the coder) and also convert digital output into analog form (the decoder). As we convert to digital switching and transmission, digital telephones are expected to replace the analog handsets.

A leading technique for digitizing voice is *pulse code modulation* (PCM), which has been used for over two decades in the telephone company's T–1 digital carriers (see below). As you speak into the handset, your voice waveform is sampled periodically, and these samples of the analog signal are digitized and transmitted. At the other end, your voice is reproduced from these digitized samples. In order for the listener to hear a good reproduction of your voice, the sampling rate has to be at least twice the highest frequency of your voice signal.[3] Therefore, for a 4000-character-per-second (4 KHz) voice channel, the sampling rate should be 8000 samples per second (which is what the T–1 carrier uses). This measures the amplitude of the analog voice signal 8000 times per second.

Each sample of the amplitude is then converted into an eight-bit number, which equals a voice digitization rate of 64,000 bits per second (64K bps).

SWITCHES

When the telephone industry was young, every telephone was wired directly into the telephone company office. Even if the other phone was in your building, you had to be connected by the operator downtown. City streets were strung with a maze of telephone wires.

The solution was to have a small version of the telephone company's switching facility in every office: a private branch exchange, or PBX. The PBX began as a manual switchboard with plugs and wires, which connected all the phones in an office without involving the telephone company. Since you could ring another telephone in your building without telephone com-

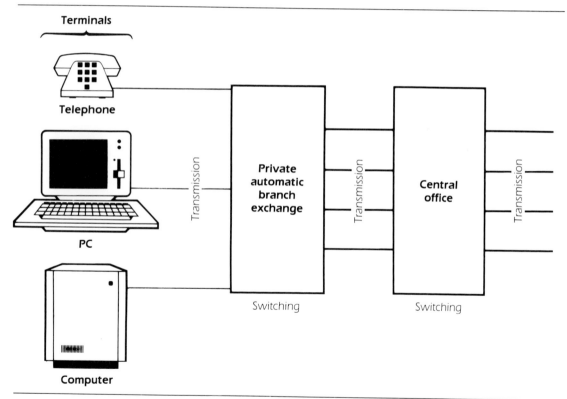

FIGURE 3-2 Telecommunications network parts.

This section on switches draws heavily on *Introduction to Local Area Networks* Order No. EB–22714–18, Chapter 8. Copyright © Digital Press/Digital Equipment Corporation (Bedford, MA), 1982.

pany switching, fewer lines were needed to the telephone office. A business with, say, 25 phones would require only five or ten outside lines.

Computer and telephone signals are basically incompatible (being digital and analog, respectively), but the telephone network is such a convenient way to communicate over long distances, it has been adapted for use with digital technology. Basically, this means converting the digital signals to analog signals via modems. Computers can thus use the telephone network of switched and leased lines as well as private branch exchanges for user-to-computer and computer-to-computer communications.

Figures 3–3, 3–4, and 3–5 illustrate three generations of switching facilities: a private branch exchange (PBX), computerized branch exchange (CBX), and digital branch exchange. A *digital branch exchange* (DBX) can switch digitized voice signals and therefore provides the option of transmitting digitized voice as well as digital data over the same line.

Transmitting voice and data over the same line has some limitations, however. The relatively low transmission speeds possible in private branch exchanges and their offspring limit their ability to switch high-speed, high-volume data traffic (such as file transfers) and to transmit high-resolution graphics. In addition, although a digital branch exchange may seem at-

---

FIGURE 3–3—————————————————————————————

PBX switches.

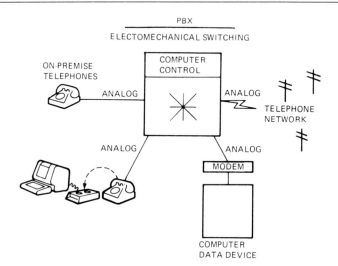

A private branch exchange (PBX) is a centralized on-site device that performs electromechanical switching. The system is all analog; internal and external telephone conversations enter the PBX directly as analog signals, and are switched as needed. Digital signals coming from a computer or terminal must be converted into analog form by a modem or acoustic coupler, and are then switched by the PBX to other computers or terminals where the signals are converted back into digital form.

FIGURE 3–4

CBX (PABX) switches.

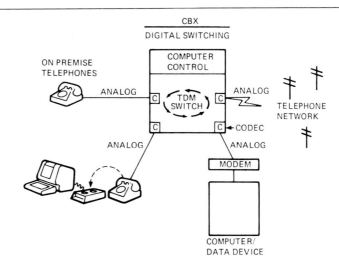

A computerized branch exchange (CBX) or priate automated branch exchange (PABX) is a computer-controlled switch that uses time-division multiplexing to allocate and switch the available channels fast and efficiently.

tractive at first glance because it uses existing phone lines, an additional wire pair is often needed to implement the new exchange, which cancels the potential benefit.

"Control of the switch means control of the customer." This saying suggests that the switch—the private branch exchange and its descendants—is the hub of the modern automated office, even though computer vendors would argue that the host processor is the natural center of any office.

TRANSMISSION

Transmitting voice digitally is both economical and reliable. Since 1962, a type of pulse code modulation called *Compounded PCM*[4] has been used to digitize voice for AT&T's T–1 carriers. The *T–1 digital carrier* is a cable with repeaters every mile or so that regenerate the signal, and has 24 voice channels *multiplexed* on the same twisted-wire pair cable instead of just a single voice channel. *Multiplexing* (see Figure 3–6) creates two or more channels out of one circuit either by splitting the frequency band it is capable of transmitting into narrower bands, each of which becomes a separate channel; or by allotting the channel to several different users in turn (this is called *time-division multiplexing*). In a T–1 cable, the 24 dig-

FIGURE 3-5———————————————————————————————

DBX switches.

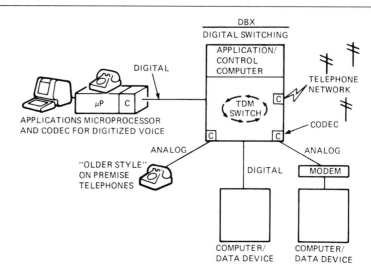

A digital branch exchange (DBX) is a digital time-division multiplexing switch. Some of the signals entering the switch are digital: digitized voices from telephones with codec chips in their handsets, digital signals coming directly from computers via private lines, and data streams sent by remote computers over long-distance digital trunk lines. Other signals are analog: voice signals from traditional telephones, and analog data signals converted from digital by modems.

itized voice channels are time-division multiplexed, which results in a bit rate of 1.544 million bits per second (24 × 8000 = 1.536 million bps, plus 8000 bps of control information). That is, 8000 *frames* are transmitted per second, each frame with 192 bits (8 × 24 bits) of digitized voice plus one framing (control) bit. The 24 digital voice channels are thus combined into one high-speed T-1 channel.[5] About two-thirds of all short- to medium-range traffic (say, between 20 and 200 miles) goes over T-1 cables.

Most transmission and switching facilities are still analog, but are rapidly moving toward digital technology. With one foot still in the analog world and the other foot already in the digital world, a typical voice transmission might follow the steps described in Figure 3-7. Depending on its route through the telephone network, the signal may be converted a half-dozen times.

————— INTEGRATED SERVICES DIGITAL NETWORKS————————

While the analog telephone network is perfectly adequate for communicating voice, telephone users and computer users have very different requirements (see Figure 3-8). *Integrated services digital networks* (ISDN)[6]

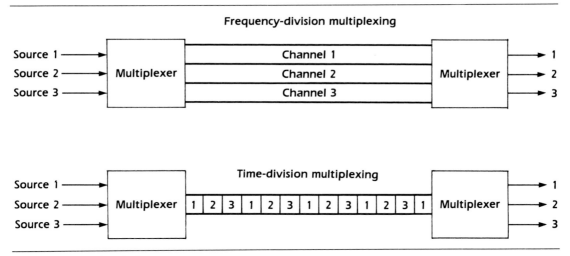

FIGURE 3–6 Multiplexing.

are an emerging solution: digital telephone networks that provide a wide array of services, including inexpensive transmission of data as well as text (e.g., electronic mail) and images (e.g., document facsimiles). In other words, using digital switching and digital transmission, ISDNs can simultaneously carry voice and data traffic on the same public telecommunications network.

ISDNs are slowly replacing the analog systems which were developed when telephones were used only for voice. And since ISDNs handle everything digitally, including voice, modems can be eliminated as well. ISDNs will not provide any services that cannot be provided by other means, but what they will do is make advanced voice, data, and image services universally available, with greater simplicity, increased reliability, and at lower cost.

These simultaneous transmissions over the telephone network are made possible by program-controlled digital switches, which are being installed by local telephone companies as well as long-distance carriers despite the many different interfaces required by the equipment already attached to the network. Standardizing these interfaces to permit universal access to an ISDN will be a critical step, and could lead to a truly global public network.

OPTIMIZATION

Generally, the volume of business telecommunications is inversely related to distance: the greater the distance, the less the traffic (see the article "Big

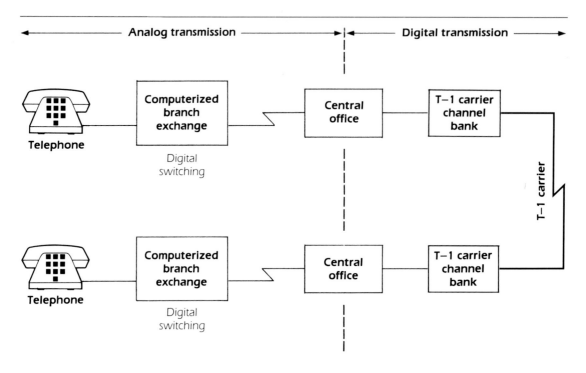

In a typical voice transmission, an analog voice signal is transmitted from the speaker's telephone to a computerized branch exchange (CBX), which converts it into digital form and switches it digitally. On the other side of the CBX, the signal is converted into analog form, and is transmitted to the telephone company's central office. The analog signal then goes to a T-1 carrier channel bank and is transmitted digitally to a channel bank at the other end, which changes it back to analog. At the receiving telephone company central office, the signal is switched in analog form and routed to the recipient's CBX. The CBX switches it digitally and sends it as an analog signal, to the telephone handset, where it is converted back into sound.

FIGURE 3-7 *Typical voice transmission.*

LAN on Campus" beginning on page 60). So far, voice and data communications have been integrated mostly for short- to medium-range traffic— between 20 and 200 miles. But voice/data integration may also be economical where data makes up a significant percentage (at least 25%) of local traffic. Two considerations should be kept in balance when choosing a network for local data and voice transmission:

■ On the one hand, a digital private branch exchange is the least expensive way to switch voice. Local area networks are not yet capable of efficiently switching voice.

■ On the other hand, digital private branch exchanges cannot now handle high-volume transmissions of data between computers. For this, local area networks using coaxial cables are usually the best solution (see "Big LAN on Campus" at the end of this chapter).

FIGURE 3–8 ————————————————————————————————————

Telephone users
vs. computer users.

TELEPHONE	COMPUTER
Telephone conversations are continuous: talking and listening.	Computer transmissions are discontinuous: bursts of data with silence between the bursts.
Noise and interference on the line are seldom critical to telephone users.	Data must travel error-free.
Switching occurs only once at the start of a phone call, and is relatively slow.	Data transmission is more efficient with fast switching of individual messages.
Telephones are simple and compatible.	Computer-related equipment tends to be complex and may not be compatible.

As all forms of information begin to move over the same network, *information managers* are needed: individuals who can provide company-wide solutions to voice and data communication problems. The traditional distinctions among voice, data, and text are no longer useful, and future data-processing managers must not only know the components of business telecommunications (e.g., front-end processors, transmission channels, multiplexers) but must also understand voice—if only because executives communicate verbally. Information has emerged as a critical corporate resource.

SUMMARY

1. Our telecommunications network is shifting from analog to digital technology as it is replaced and expanded.

2. Since the 1970s, voice, data, text, and image have all been converging, and they can all be transmitted digitally.

3. The telecommunications network has three major parts: terminals, switches, and transmission facilities. Each part is being replaced with digital technology at different rates.

4. A leading technique for digitizing voice transmission is pulse code modulation, which has been used for over twenty years in the telephone company's T–1 carriers.

5. Computers and related devices use digital technology.

6. The office private branch exchange is the hub of the modern automated office.

7. Ultimately, all switching and transmission will be digital, and voice will be in analog form only at either end.

8. Integrated services digital networks are digital telephone networks which transmit everything (i.e., voice, data, text, and image) in digital form and on the same channel.

REVIEW QUESTIONS

1. Explain analog and digital transmission.

2. What are the three major parts of our telecommunications network?

3. Codecs are analogous to modems. Explain.

4. Define: pulse code modulation, T–1, 4KHz.

5. Why are we moving toward digital transmission?

6. Explain why the private branch exchange developed.

7. Describe how a computer branch exchange (CBX) works.

8. A digital branch exchange is a time-division multiplexing switch. Explain.

9. "Control of the switch means control of the customer." Why?

10. Explain why a signal is converted as it is routed through the telephone network.

11. Comment on the differences between voice conversations and data transmission.

12. Why are we moving toward integrated services digital networks (ISDNs)?

13. The article at the end of this chapter says LANs save fuel and power. Explain.

ENDNOTES

1. See U. D. Black, *Data Communications, Networks and Distributed Processing* (Reston, VA: Reston Publishing Co., 1983), 209–210, for a discussion of the advantages of digital transmission.

2. See George A. Trimble, Jr., *Digital PABX* (Princeton: Carnegie Press, 1983), Chapter 3, for a discussion of terminal digitization.

3. This is the Nyquist Theorem. See A. S. Tanenbaum, *Computer Networks* (Englewood Cliffs: Prentice–Hall, 1981), 95.

4. Companding (compressing or expanding) the analog signal gives preferential treatment to the weaker parts of the signal by imparting more gain to weak signals than to strong signals. Low-level signals are boosted and the larger amplitudes are held constant. The reverse occurs at output. See James Martin, *Future Developments in Telecommunications* (Englewood Cliffs: Prentice–Hall, 1977, 2nd Ed.), 525.

5. Digital streams can be combined and transmitted at four basic speeds:
   T–1 = 1.544 million bps (twisted-wire pair, 24 voice channels)
   T–2 = 6.312 million bps (twisted-wire pair, 96 voice channels)
   T–3 = 44.736 million bps (fiber optic cable, 672 voice channels)
   T–4 = 274.176 million bps (coaxial cable, 4032 voice channels)
   See James Martin, *Future Developments in Telecommunications* (Englewood Cliffs: Prentice–Hall, 1977, 2nd Ed.), 501.

6. See R. A. Miller, "ISDN and the Information Age: Real World Implications," *Telecommunications,* November 1985, 64a.
7. See "Telecommunications Survey," *The Economist,* November 29, 1985, 7.

BIG LAN ON CAMPUS

Brown University, Providence, RI, has become one of the first major U.S. universities to install a major local area network on its campus. The Brown system, designated BRUNET, is based on the use of CATV components and a large Sytek LocalNet 20 system. The Brown system will ultimately interconnect 110 buildings and provide more than 1,000 connections. The long-term goal is to connect BRUNET to all offices, lecture rooms, laboratories, and dorms.

The host devices connected via BRUNET include DEC VAX 11–780s and IBM 4341s and IBM 370–158/3 processors. The network provides concurrent 9600 bit/second connections for hundreds of ASCII terminals. In addition, it provides dial-in services for non-connected users and includes a gateway connection for access to external services. BRUNET also handles a campus-wide energy management system, a campus security system, and television transmission.

Installation began in 1981. The cost of the installation of the primary, or backbone, LAN distribution system was $317,000. The total cost of additional items such as service equipment, 1,600 building wiring outlets, security systems, and other items, was an additional $1.024 million. In reviewing BRUNET's overall installation costs,

one must consider the system spans about 15 miles, considerable effort was spent on specialized applications, and its installation required some complex efforts—such as wiring the University Hall built in 1764.

The payback period of the LAN was partially based on an expected annual fuel and power cost reduction of $600,000 through the LAN's energy management. The annual LAN operation costs are expected to be about $150,000—mostly labor. The monthly LAN service charge per user is $20, which includes all maintenance. That is less than half of the charge previously paid for telephone-based connections.

As an extreme example of LAN serviceability, a fire in the distribution tunnel system caused by a telephone serviceman destroyed 2,600 pairs of telephone wires and all data cables. Due to the flexibility of the network, BRUNET was operable about 40 minutes after the fire was extinguished. Although this was accomplished by temporary cables and emergency repair efforts, the other types of communication services took up to several days to restore.

Currently, user demand for access to BRUNET is growing at about 30% to 40% per year.

Reprinted from the July issue of *Modern Office Technology*, and copyrighted 1984 by Penton/ IPC, subsidiary of Pittway Corporation.

4 SOME BUSINESS TELECOMMUNICATIONS COMPONENTS

NETWORKS

TRANSMISSION CHANNELS

MULTIPLEXERS AND
    CONCENTRATORS

FRONT-END PROCESSORS

PORTS

SWITCHING

PACKET-SWITCHING
    NETWORKS

PRIVATE NETWORKS

AFTER READING THIS CHAPTER, YOU WILL UNDERSTAND THAT:

- Business telecommunications components often include transmission channels, multiplexers, concentrators, front-end processors, and ports.
- Switching is an important network function.
- Public packet-switching networks are to data transmission what the telephone system is to voice.
- Private local and long-distance networks are attractive to many organizations.

Computers and communications are part of the same continuum. Wang, for example, has equipment compatibility agreements with over a dozen private branch exchange manufacturers and an interest in U.S. Satellite Systems. ICL, Britain's oldest and largest computer company, was acquired by a British telecommunication company, STC. In marketing terms, control of communications services can increase a company's sales of computers and related equipment.

This chapter discusses some typical components of business telecommunications systems, beginning with networks.

NETWORKS

Networks are a familiar concept: highways, television stations, and rivers are all networks of a kind. Telecommunication networks are also familiar: AT&T and MCI, for instance. In telecommunication networks, a physical circuit between two points is called a **link** or **channel,** and a junction of network links is called a **node;** so a telecommunications network consists of links and interconnected nodes.

A network has three parts: a transmission system, switching devices, and stations. A station for a voice communication network is a telephone; in a data network, stations may be telephones, terminals, or facsimile machines. Switching ensures that each station can connect to every other station.

Networks can offer two important benefits:

- Hardware and software resources can be shared, which reduces outlay for printers, modems, and software packages.
- Information can be shared, which helps decision making, provides greater flexibility, and improves responses. Networks can help get the right information to the right person at the right time to make the right decision.

The most familiar type of network is probably the private branch exchange, which was developed to handle analog voice signals. Currently, private branch exchange systems either combine digital computer data with analog voice signals, or they convert all of the signals into digital form. Private branch exchanges readily support functions such as sharing and manipulating spreadsheets and transferring data between personal computers. Using traditional twisted-wire pairs, however, there are limits to speed and capacity.

Ethernet is a local network jointly developed by Xerox, DEC, and Intel which employs a baseband coaxial cable to transfer digital signals among its nodes—all the devices attached to it. Baseband systems encode and modulate data signals so that at any point on the link or channel only one signal at a time is present (this is also called time-division multiplexing). The usual medium for the link is a coaxial cable, a flexible single conductor also used for cable television. Best results are obtained when the transmission distance does not exceed two miles and the attached equipment shares operating protocols such as transmission speed. Ethernet is a good way to share the information in disk drives and data bases, but its baseband technology is not designed to handle nondigital signals such as voice and video.

Sharing data-storage devices and other hardware via these networks can be economically attractive: installing a network can be less costly than multiple printers and data bases. On the other hand, most software packages and operating systems are currently designed for single users, although multiuser versions are under development.

There is already a growing range of networking options available, and as prices fall, more organizations will be considering telecommunications. The emphasis should be on the firm's requirements: information volume, compatibility, and growth. For example, organizations which need to transmit a large volume of voice and data over distances longer than 500 miles may benefit from microwave and satellite technologies. For local requirements, the building itself may be the solution: heating, ventilation, air conditioning, security, elevators, and lighting can all be controlled by a central computer. Newer buildings may also offer shared telecommunications and computer networking facilities.

Remote computing is another networking option. Personnel in the field can transmit messages and manipulate data base information from a portable terminal or computer connected to a stationary mainframe. For example, a salesperson can check on the status of a customer's order, or a maintenance engineer can learn if a part is in stock. All you have to do is match protocols so the mainframe computer can recognize, accept, and process calls from the mobile unit. Protocols, as we discussed in Chapter 2, are software formats that establish bit, character, or message synchronization between communicating devices, enabling the devices to read data from each other as well as recognize and correct errors.

——— TRANSMISSION CHANNELS ———

Information travels from one point to another along a transmission link (also called a channel, trunk line, or circuit). The link may be made of wire, coaxial cable, microwave signals, optical fibers, or any other medium. An intercity telephone call, for example, might travel over all of the following links:[1]

LINK	ROUTE	MEDIUM
1	Telephone to local telephone company central office	Twisted-wire pair cable
2	Local central office to toll switching office	Twisted-wire pair cable, coaxial cable, or fiber-optic cable
3	Toll switching office to toll switchingoffice	Coaxial cable, microwave transmission satellite relay, or fiber-optic cable
4	Toll switching office to local central office	Twisted-wire pair cable, coaxial cable, or fiber-optic cable
5	Local central office to telephone	Twisted-wire pair cable

The earliest medium for telephone and telegraph transmission was a pair of wires, each pair being one telephone channel. These pairs were largely replaced by cables which contained many channels (twisted-wire pairs), and later by coaxial cables which have little distortion or signal loss. A coaxial cable two inches in diameter can handle about 1,000 simultaneous conversations. Obviously, cables may be either strung above ground or buried, both of which usually require a right of way.

Fiber optics are hard to beat since transmission is rapid, accurate, and almost immune to interference. Fiber optics use pulses of light transmitted via hair-thin strands of glass. The light source is either a light-emitting diode (LED) or a laser, and the beam of light can be modulated to carry huge amounts of information. Britain has fiber-optic systems that use one optical fiber to carry 2,000 simultaneous telephone conversations on a single light beam at a rate of 140 million bits per second. At present, fiber-optic cables are difficult to connect since they have to be aligned with microscopic precision before they are glued with epoxy or fused together. Also, light signals must be converted into electrical current for switching.
Fiber-optic cables have several advantages:

- low error rates
- glass is made from silica (sand), which is cheap and plentiful
- huge capacity for carrying information (up to 140 million bits per second)
- low susceptibility to electrical interference
- smaller and lighter than wire cables

Microwave signals avoid some of the legal, construction, and maintenance problems of cables. Thousands of voice channels can be carried without any physical connections between the microwave relay towers. The signals follow a straight line and the relay towers have to be in line of sight with each other (usually twenty to thirty miles apart). Each tower receives the signals, amplifies them, and retransmits the signals to the next tower. Most long-distance telephone links are now either coaxial cable or microwave transmissions.

Satellites are like very high microwave relay towers that use radio waves instead of microwaves. They link two or more earth stations which transmit on one frequency and receive on another frequency so the incoming and outgoing signals do not interfere with each other. The satellite receives the transmission, amplifies or repeats the signal, and retransmits it on the second frequency. These different frequency bands are called transponder channels, since a *transponder* inside the satellite receives the transmissions from earth, amplifies them, changes their frequency, and retransmits them back to earth. To keep the satellite within line of sight of the earth stations, it remains in geosynchronous orbit 22,300 miles above the earth, exactly matching the earth's period of rotation. Satellites can handle large volumes of voice and data transmissions simultaneously, and satellite transmission services are offered by MCI, Communication Satellite Corp., Western Union, and others.

MULTIPLEXERS AND CONCENTRATORS

If the same number of telecommunication devices can be supported with fewer lines, this will cut costs. One way to do this is by multiplexing—combining several independent channels into one high-speed data line, thereby reducing the telephone bill. Figure 4–1 shows how data streams from three terminals can be merged into a single data stream, which means that only one line has to be leased.

Synchronous time-division multiplexing divides the data stream between the multiplexers into *cycles*, and every cycle has a time slot for each terminal. This is not optional; there must be a slot for each terminal, even when a terminal is idle. Clearly, this wastes some transmission time, since the slots in the data stream for idle terminals will be empty.

Low-speed data streams from several terminals go into a multiplexer, which combines them and retransmits them at high speed. Another multiplexer "breaks apart" the data and reconstitutes the original low-speed data streams. Synchronous time-division multiplexing can take place at either the bit level or the character level.

Asynchronous time-division multiplexing (commonly called statistical multiplexing) makes better use of the high-speed channel, accommodating three or four times as much traffic on long-haul leased lines. The time slots are allocated only among the *active* terminals, so there are no unused time

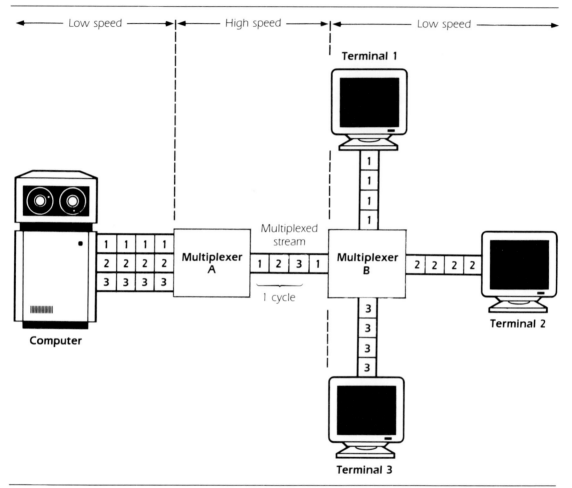

FIGURE 4–1 *Synchronous time-division multiplexing.*

slots in the high-speed data stream. The device that does this is called a **concentrator,** statistical multiplexer, or communications processor. A concentrator can do more than a multiplexer: it can temporarily store data, it can alter the form of the data stream, and it can even be programmed.

Since the concentrator is a funnel for all the data that passes through the network, it is a logical place to make code, speed, or format conversions. Any variations in transmission speed, codes, format, or line protocol may make a terminal incompatible with a central computer. But by having the concentrator do any necessary conversions, we can make any terminal compatible with the mainframe, presenting data and messages in the form the mainframe expects.

————FRONT-END PROCESSORS————

Communicating with many local and remote terminals can put a heavy load on a central computer. Often a mainframe is acquired to run application programs and retrieve data; not to handle a data communications network. To make things easier on the mainframe, another computer—a front-end processor (FEP)—can take over the network housekeeping chores, allowing the mainframe to use its valuable processing time for its intended tasks. For instance, the front-end processor for a DEC 20 system might be a DEC PDP-11, and for an IBM system it could be a 3705 or 3725 Communications Processor. Figure 4-2 illustrates the FEP, multiplexers, modems, high-speed lines, and terminals that might be installed to make optimal use of a mainframe.

Since the front-end processor controls all communications between the network and the mainframe, it is a critical component in a telecommunications network. The heart of the FEP is the *network control program (NCP)*, software written in macro language that instructs the FEP how to talk to both the central computer and the terminals. Its functions include the following:

- **Translating codes.** A network may have terminals that use different codes (e.g., ASCII, EBCDIC), so the NCP translates all input using these codes into the internal code used by the mainframe, and vice versa.

- **Managing multidrop lines.** Several terminals can share a single multiterminal or **multidrop line** to the mainframe (see Figure 6-3 on page 99). The NCP can query each terminal on the line for messages, address specific terminals in turn, and then direct the computer's answers to the appropriate terminals. The NCP also monitors the lines, automatically removing terminals that don't answer and entering those that transmit errors on an error report.

- **Handling traffic.** The NCP handles two sets of buffers: one that queues inquiries intended for the mainframe, and another that queues the replies going back to the network from each port of the FEP (ports are discussed in the next section).

- **Inserting and deleting control characters.** Just as an address is needed to get a letter to its destination, address and control characters are needed by every message travelling over the network. The NCP attaches these characters to outgoing messages, and strips them off incoming messages so that the central computer will receive only the actual message.

- **Converting formats.** The mainframe and its peripherals exchange data in parallel format (one character per cycle). The network uses serial format (one bit per cycle). The NCP converts from one format to the other.

FIGURE 4–2 ——————————————————————————
Front-end
processor (FEP).

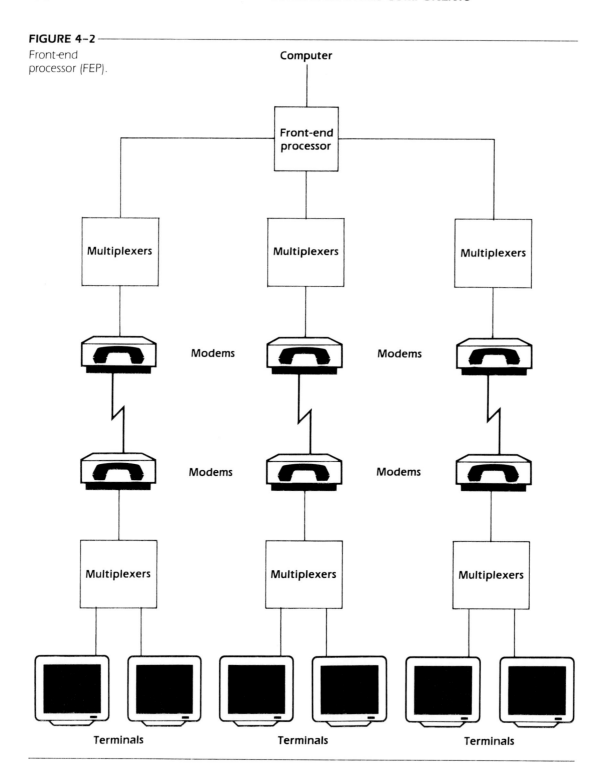

■ *Recording network statistics.* The NCP can monitor network performance, collect transmission load data, and display diagnostic messages that alert the operators to any problems.

PORTS

Terminals are connected to the computer (or more precisely, to the front-end processor) through ports. A port is a physical location for data entry or exit; it is the part that receives data from or transmits data to one or more external devices over a single data channel.

Computer ports are expensive, and are usually installed in groups of eight or 16. Since the number of terminals a computer or front-end processor can service depends on the number of ports, they can be a limiting factor in setting up a telecommunications network. Port-sharing devices (see Figure 4–3) can be used to attach more than one terminal to a port, but only one terminal can transmit at a time.

SWITCHING

Dialing the telephone number (617) 999–8605 automatically invokes the switching capabilities of the telephone network. The area code—617—connects you to the correct toll switching office, and the next three digits, 999, connect you via the toll trunk line to the appropriate telephone company central office. The last four digits, 8605, then connect you through the local loop to the telephone you are calling.

This example illustrates how important switching is in many telecommunications systems. There are three general types of switching:

■ *Line-switched systems* (also called circuit-switched systems), which are similar to the telephone network. The routing is set up before the message is transmitted, and the circuit is tied up for the length of the transmission.

■ *Message-switched systems,* which were developed to offset the errors that sometimes occur when analog lines are used for digital messages. Messages are stored in a buffer until the transmission is complete, and the whole message then goes to a concentrator where fields are added for identification, address, and error detection. The message with these added fields is then sent down the line to its destination. With message-switched systems, no circuits are tied up; instead, the message is passed through the network from node to node, and is stored briefly at each node before being retransmitted. While message switching is generally good for digital data exchange, such as electronic mail and swapping computer files, its long and highly variable delays make it unsuitable for interactive applications.

FIGURE 4-3

Ports.

- **Packet-switched systems** (see below) are a type of message-switched system which divide messages into blocks or packets of equal length, each carrying its own fields for identification, address, and error detection. Each packet can then travel a different route, reducing congestion.

Line-switched and message-switched systems (including a packet-switched systems) make up a classification called *switched communication networks*. These networks consist of a group of interconnected nodes, and data is switched from node to node as it is routed through the network from source to destination.

FIGURE 4–3———————————————————————————————

continued

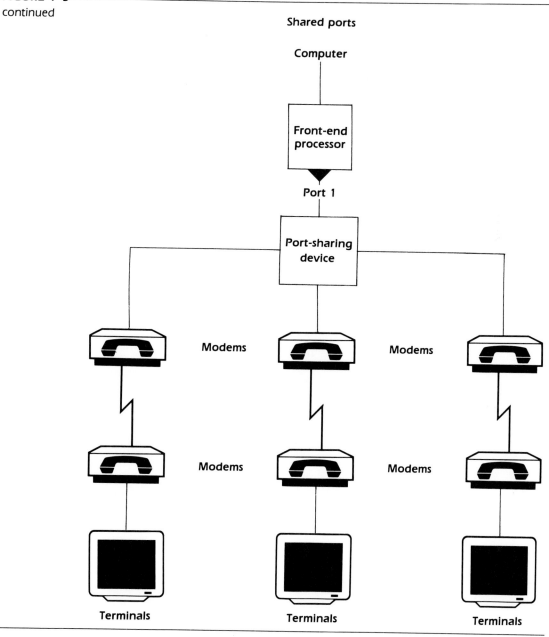

In contrast, a *broadcast communication network* has no intermediate switching nodes. Rather, each station is connected to a transmitter/receiver which communicates directly with all the other stations. In CB radio, for example, all stations tuned to the same channel can communicate, although only one station can transmit at a time. For this reason, a broadcast communication network often calls for controlled access.

PACKET-SWITCHING NETWORKS

Public packet-switching networks were developed to carry digital data, much as the telephone network was designed to transmit voice.

In a packet-switching network, each digital data message is divided into standardized blocks or *packets*. Each packet contains a destination address, an origin address, control information, and the actual data being communicated. Several packets may be needed for a long message, and since each packet may be routed differently, they might arrive out of sequence. The control information allows them to be reassembled in proper order.

Figure 4–4 illustrates a packet containing up to 128 bytes, or characters, of data. Note that the data itself has a header. This packet is then enclosed in a *frame,* another packet that contains a header as well as a trailer. As shown in Figure 4–5, the frames are then sent to the packet-switched network.

The Consultative Committee on International Telegraphy and Telephone (CCITT) has recommended a worldwide standard—called **X.25**—for linking equipment to public packet-switching networks. No changes are required, for example, to link a computer to a packet-switching network, since a software package containing the X.25 protocols can convert the computer's output to the network standards and convert the network's output so the computer can deal with it. This software package may reside in the computer or in the communications controller.

Telenet, a subsidiary of General Telephone and Electronics Corp., is a leader among packet-switching networks. The cost of using Telenet is independent of distance, and depends on the volume of data packets sent and the time spent connected to the network. Other packet-switching services used for data communications include Tymnet, ITT, and AT&T Information Systems.

PRIVATE NETWORKS

In addition to public networks for voice and data communications, there are also private networks. A subscriber who owns or leases a private line or network of lines will never get a busy signal because the line is always available. Private lines can be conditioned by adding electronic components to improve the quality of transmission (e.g., fewer errors and higher speeds).

A good example of a leased voice line is foreign exchange service. Let's say that a downtown store has many customers who live in the suburbs,

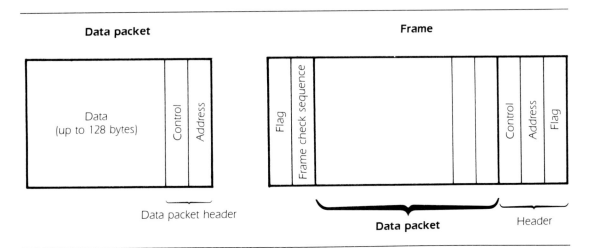

FIGURE 4-4 How packets are assembled.

in a different telephone exchange. It may benefit the store to lease a private line that connects the store (or its private branch exchange) to the suburban telephone exchange. Suburban customers could thus avoid long-distance charges when calling the store downtown.

Local area networks (LANs) are a different kind of private network. Hospitals, colleges, offices, and manufacturing plants all generate a lot of

FIGURE 4-5 ————————————————————————
Packet assembler/
disassembler (PAD).

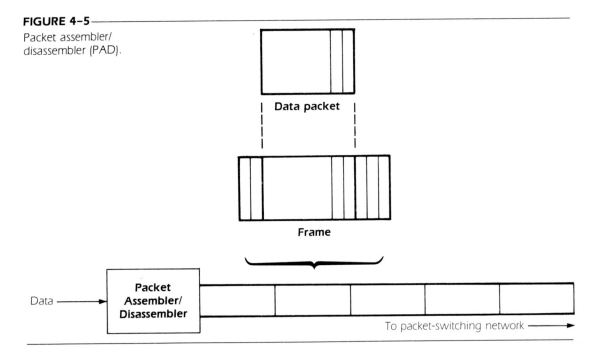

information that is distributed and used locally, and LANs take advantage of this pattern by connecting personal computers, word processors, disk storage units, mainframes, and the like (see the article "Market Trends: Satisfied Customers" beginning on page 77). LANs seldom extend for more than a couple of miles and are self-contained, even when connected to a common carrier.

An essential component of any network—transmission channels, multiplexers, concentrators, front-end processors, ports, switching, packet-switching networks, and private networks—is software. Software programs supervise, manage, control, and otherwise make diverse pieces of equipment work together in an efficient, orderly manner. In Chapter 5, we will review some of the different types of software, and look at the functions that software performs as well as where software instructions reside.

THE BRISFIELD COMPANY

The Brisfield Company's distribution manager (see end of Chapter 1) has asked you to recommend improvements in the system described in the *Inventory Control* section. For example, she wants to reduce her telephone bill and to cut the time her staff spends keeping inventory records current. What would you recommend to modernize this system?

SUMMARY

1. A telecommunications network consists of a transmission system, switching devices, and stations.

2. Transmission channels may be wires, cables, microwave signals, or optical fibers.

3. Multiplexers and concentrators reduce the number of lines needed by merging several streams of data into one stream and then breaking them out again at the other end.

4. Front-end processors are computers that do the network housekeeping chores, saving the mainframe's processing time for other tasks.

5. Computer ports control computer access and are often a limiting factor.

6. In a line-switched system, routing occurs before message transmission and the circuit is tied up for the duration of transmission.

7. In a message-switched system, the message is passed from node to node to its destination. Message switching is good for electronic mail and computer files. Packet switching is a type of message switching.

8. Public packet-switching networks were developed to carry digital data.

9. The standard for linking equipment to a packet-switching network is CCITT's X.25 protocol.

1. What is packet switching?
2. What is the X.25 protocol?
3. Describe foreign exchange service.
4. Comment on local area networks.
5. Describe the various transmission channels.
6. How do satellites fit into business telecommunications? Optical fibers?
7. Define and explain multiplexer and concentrator.
8. What is the function of a front-end processor? Port?
9. Explain the three general types of switching.
10. The article at the end of the chapter explains why some businesses turn to local area networks and points out some installation snags. Discuss.

ENDNOTE

1. See W. John Blyth and Mary M. Blyth, *Telecommunications: Concepts, Development and Management* (Indianapolis: Bobbs–Merrill, 1985), 103.

MARKET TRENDS
Satisfied Customers

HARRY HENRY

A survey of 265 communications managers at U.S. user organizations found a large installed base of micros waiting to be networked. User attitudes seem to be: "Now that we went on a spending spree and bought all these micros, let's do something with them before we buy any more."

Information systems managers are now thinking more economically about how better to utilize current resources to satisfy end-user demands. The local area network solution is popular—40% of respondents currently have installed a network, and another 29% answer that they are planning to do so.

Results of the survey, conducted by the Market Information Center of Framingham, Mass., peg the following as the top three reasons for implementing a local area network:

- to share the data base
- to share hardware resources
- to see a return on investment

MEASURES OF SATISFACTION

	NUMBER	%
Did it meet expectations (after installation)?		
Yes	73	74.5
No	9	9.2
Undecided	16	16.3
Total	98	100.0
Would you recommend network product?		
Yes	71	73.2
No	6	6.2
Not sure	20	20.6
Total	97	100.0

Source: The Market Information Center, Inc.

Figure 2

ASSESSING NETWORK PURCHASE

REASONS FOR IMPLEMENTATION	POINT VALUE
Share data base	111
Share resources	89
Economy	51
Increase system speed	34
Ability to move workplaces	32
Connect differing devices	31
Mainframe link	30
Expansion	24
Increase productivity	24
More efficient operation	24
Link micro to micro	19
Users more access	18
Electronic mail	16
Reduce CPU overhead	13
Technology became available	12

Source: The Market Information Center, Inc.

Figure 1

INSTALLATION SNAGS

SINGLE BIGGEST PROBLEM	NUMBER OF MENTIONS
Software support	10
Cabling	7
Resource limits	5
Getting going	4
Hardware incompatibility	4
Having users use it	4
Weak communications support	4
Disk problems	3
Price	3
Vendor support	3
Application programs	2
Diagnostics	2
Moving users	2
Multiple vendors	2
No time to learn	2
Printer support	2

Source: The Market Information Center, Inc.

Figure 3

From *Computerworld*, September 25, 1986, p. 41. Reprinted with permission of The Market Information Center, Marlborough, MA.

These issues do not involve productivity, standards, or stringing cable; they are practical issues (Figure 1). Users are looking for products that can help them manage their current resources more effectively.

The satisfaction level of current users with their networks is high—74% indicate their network met their expectations after installation (Figure 2). Apparently the user community is finding solutions to networking problems with current products. Vendors may not agree on standards, but users report the products are working and serving their purposes.

This finding may surprise market watchers observing fairly slow growth rates. Because relatively few companies have installed local area networks, it appears there must be something technologically wrong with the networks. This negative attitude probably stems from the installation stories. Nearly every company suffers installation problems, however minor, and these problems continue to be reported.

The four major problems users encounter most often are software support, cabling, resource limitations of the local area network, and, finally, the user inertia factor (Figure 3). Yet the same high percentage—73% of those surveyed—would recommend their systems to other users.

Users are actively looking for the IBM answer. Less than 10% of companies with networks installed indicate IBM as the primary vendor, but among companies planning to install a local area network, more than 20% will consider IBM in their product decision (twice as many as the next vendor). Other vendors being considered include Ungermann–Bass, Inc., Digital Equipment Corp., 3Com Corp., and AT&T, with Sytek, Inc., Novell, Inc., Corvus Systems, Inc., and 3M Co. not far behind.

The two major suppliers of installed systems are DEC (with the Ethernet influence) and Sytek (with an IBM tie-in). Following them are IBM, Wang Laboratories, Inc., and others. Nine percent of users develop network solutions internally.

While the vendors and standards committees continue to sort out issues involved in network technology and interfaces, customers must deal with choices regarding wiring costs, protocols, and cabling media. Their buying decisions will make users the final judges of the products' success and utility.

Judging by outward appearances, communications managers are managing the business of local area networks quite nicely. Systems are being installed, end users appear satisfied, and the market (at least so far) is being driven by the need to use resources effectively. This purchasing pattern contrasts with the "micro fever" of the past few years, when many micros were purchased without careful attention to how they would be used. As the market matures, so do the products within it. The users are exercising their considerable influence to good effect.

5 SOFTWARE

INTRODUCTION
OPERATING SYSTEMS
TELEPROCESSING MONITORS
DATA BASES
NETWORKING SOFTWARE

LINE CONTROL SOFTWARE
TELEPROCESSING SOFTWARE
PERFORMANCE SOFTWARE
APPLICATION PROGRAMS

AFTER READING THIS CHAPTER, YOU WILL UNDERSTAND THAT:

- A computer's operating system has a lot to do with its telecommunications capabilities.

- There usually are two software layers between the operating system and application programs: telecommunications access software and teleprocessing monitor software.

- Telecommunications access software helps application programs access the network and move data to and from the mainframe.

- Teleprocessing monitor software provides application programs with a standard interface to the telecommunications access software, and relieves the application programs of tasks such as line control and data recovery.

- Data bases are important in telecommunications because they contain the information remote users need. Data bases are maintained by data base management systems.

- The important elements of networking software are: line control, teleprocessing control, performance monitoring, and the applications themselves.

INTRODUCTION

If hardware is the computer itself (and any related equipment), software is the programs and routines (along with their documentation and operating procedures) that operate the computer and instruct it what to do. These programs may be scattered throughout a network; for example, remote personal computers would have their own application programs (e.g., dBase III Plus, Lotus 1-2-3, Microsoft Word), the **cluster controller** would have software to help control data flow to and from the terminals, and statistical multiplexers at each end of the line might use switching, store-and-forward, and control software. The front-end processor would use a software program for network control, and the mainframe would normally have an operating system, a teleprocessing monitor, a data base management system, and various application programs (see Figure 5–1).

Software functions are increasingly found in microchips instead of magnetic tape, floppy disks, or internal storage devices. This so-called **firmware,** usually faster and more reliable than other storage media, tends to blur the distinction between hardware and software. To make things more confusing, some protocols are embedded in software such as telecommunications across programs.

This book often refers to IBM operating systems, protocols, hardware, and related devices. This is not an endorsement of IBM products, but since many hardware and software vendors strive to be compatible with IBM, it is a practical way to illustrate the concepts.

OPERATING SYSTEMS

Figure 5–2 shows the main memory of a typical mainframe or host computer which provides services such as computation, data base access, and programming languages to a network. The main memory holds the **operating system,** * a collection of software instructions that tell the computer how to communicate with remote devices and read other software programs, as well as the **application programs** for specific uses such as processing payroll or controlling inventory.

As shown in Figure 5–3, the operating system actually consists of three kinds of programs:

- *language translators,* such as COBOL and FORTRAN compilers, which facilitate the writing of application programs
- *support programs,* including diagnostic aids, which help keep the system working smoothly

*Operating systems are also called master control programs or system software.

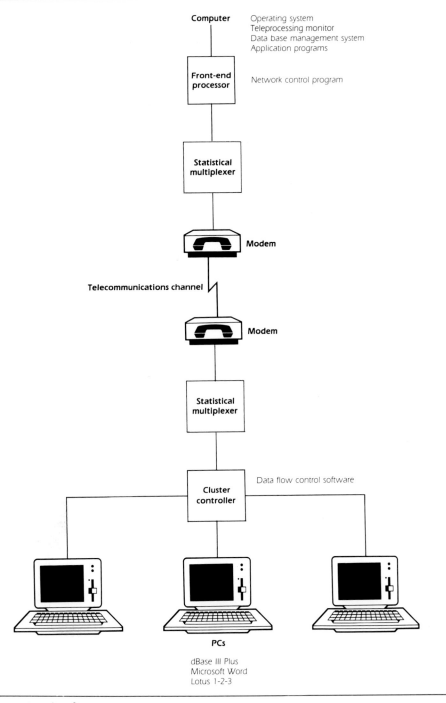

FIGURE 5-1 Some network software.

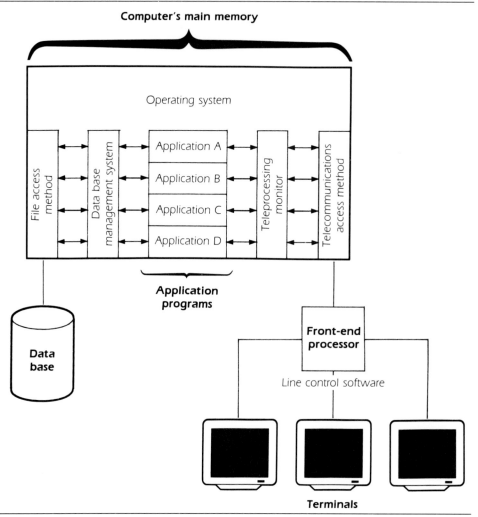

FIGURE 5–2 Telecommunications software.

■ *control programs,* which coordinate and schedule the running of application programs, handle input/output operations, process interrupts, and deal with system errors.

The operating system thus helps to manage all the input, output, and processing functions in the system. For example, it will allocate main memory to the various application programs, and coordinate the tape and disk drives with the central processing unit. The operating system's *message handling* or telecommunications functions consist of the following:

Job management refers to how the operating system handles requests for processing a stream of jobs coming from peripheral devices and remote sites. It places these incoming jobs in a file called the *input queue*, and as resources become available, the operating system's job management will choose the next job to be executed based on length of time in the queue, job priority, availability of peripheral devices, memory storage space required, and so on. Job management then selects the next job, brings it into main memory, and turns it over to task management.

Task management allocates the computer's resources to specific tasks, such as accessing files and processing segments. It assigns memory storage space to these tasks and schedules their execution, quite a juggling feat when you consider that several programs may be running at the same time in a multiprogram system. Finished jobs are put in an *output queue*, which is handled by job management.

Data management provides input/output services for application programs, such as allocating space, accessing files, and blocking records. Data management also handles input/output for telecommunications links as well as tape and disk drives. In this book, we are mostly interested in the input/output operations associated with data transmission, such as the telecommunications access methods discussed later in this chapter.

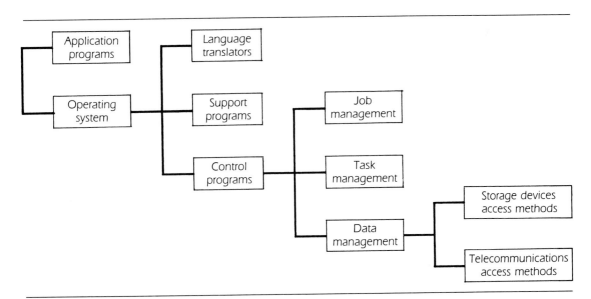

FIGURE 5–3   Operating system structure.

Adapted from James Martin, **Introduction to Teleprocessing**, © 1972, p. 165. Reprinted by permission of Prentice-Hall, Inc., Englewood Cliffs, New Jersey.

TELEPROCESSING MONITORS

Telecommunications access programs are the interface between the operating system and the telecommunications hardware, and allow application programs access to the network so data can be moved in and out of the host computer. *Teleprocessing monitors* are software programs that provide the application programs with a standard interface to the telecommunications access method software described later in this chapter (see also Figure 5–2). The teleprocessing monitor relieves the application program of line control tasks (e.g., connecting, disconnecting, flagging errors, retransmission) and other electronic drudgery.

IBM's Customer Information Control System (CICS) is a teleprocessing monitor that allows transactions from remote terminals to be processed by more than one application program at a time. CICS starts and stops tasks, decides which task runs first, provides the interface between application programs and peripheral devices, and even supports the use of data bases. Thus teleprocessing monitors interface not only with the telecommunications access method software, but also with the operating system, application programs, and other software packages such as data base management systems.

DATA BASES

Point-of-sale terminals at the check-out counters of supermarkets read the bar codes on products as the checker passes them over the optical reader. The bar code data then goes through a multiplexer, which identifies the terminal, and over a twisted-wire pair cable to a minicomputer in the back of the store. The product and package size are identified by the computer, which then retrieves the price, computes any sales taxes, and sends the total back through the cable to the terminal. The computer stores a record of the transaction, and automatically updates inventory records as well (see Figure 5–4).

A **data base** is a collection of data organized to avoid duplication and permit retrieval of information for a wide variety of uses; in this case, the data base is a constantly updated electronic file containing product prices, sales volume, inventory information, and the like. A **data base management system** (DBMS) is a software program that handles the organization, cataloging, storage, retrieval, and maintenance of data in a data base.

Figure 5–2 shows how the DBMS provides application programs with access to the data bases. The DBMS interfaces with the application programs, the operating systems, and the data bases, as well as with other software programs such as the file access method software.

In short, the software is at least as important as the hardware. Software provides much of the order and control in a telecommunications system, even when some of the functions overlap untidily as in Figure 5–2.

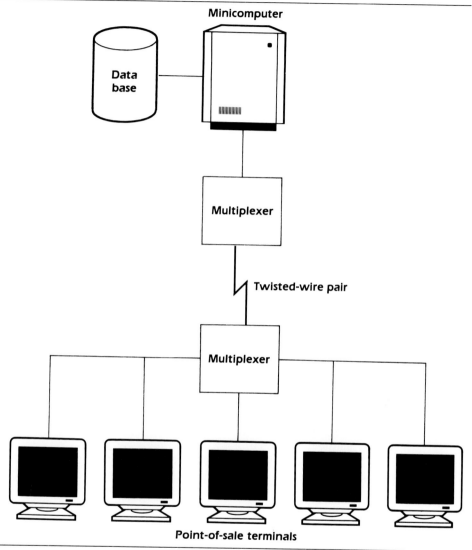

FIGURE 5-4 Data base application.

NETWORKING SOFTWARE

Hardware plus lines and access to a common carrier does not equal a network. Almost every component of a network is controlled by software, and it is these software programs that make the network function as a unit. Compatibility is crucial: the network's effectiveness is determined by how well these programs work together (see "Spencer Gifts Takes Credit a Step Farther" at the end of this chapter).

Networking programs can be categorized as follows:

- *Line control software* physically operates the network. The network control program that runs on the front-end processor is an example.
- *Teleprocessing software* resides in the central processing unit and controls exchanges between the CPU and the network. Telecommunications access method programs fall into this category.
- *Performance software,* such as IBM's Network Performance Analyzer, provides network status and diagnostics.
- *Application programs* reside in the CPU and provide specialized services to users.

LINE CONTROL SOFTWARE

Line control software operates the network, and since this usually involves hardware and software from a variety of vendors, common standards are needed in the following areas:

- *device control,* such as tabbing or going to a new page
- *data transmission,* such as initiating, checking, and retransmitting data between two devices
- *data formats,* including data structure and control characters

The protocols governing these matters will be discussed in the next chapter.

Line control software usually runs on a front-end processor, which relieves the central processing unit of the burden of controlling the network. As noted in Chapter 4, this includes functions such as line control, code conversion, error recovery, and stripping incoming messages of their control characters.

An example is the network control program (NCP), which resides in the front-end processor and talks to the CPU on one side and the network on the other. In order to operate the network, the NCP needs to know the characteristics of all the devices, channels, and lines connected to the network, as well as how they are interconnected. Accordingly, the network control program uses a *network definition data base* to do its job. This data base contains the following information:

- *Host.* Describes the number and size of the buffers to allocate to the channels that move data in and out of the mainframe.
- *Devices.* Specifies device characteristics, including type of device, network address, and buffer size.

- **Lines.** Describes each transmission line, including line speed, polled or unpolled, full-duplex or half-duplex, and modem protocol.
- **Line groups.** Specifies whether lines are leased or dial-up, as well as line control protocol.

Whenever the network is changed — for instance, if equipment is added or removed — the network definition data base must be updated as well.

TELEPROCESSING SOFTWARE

Teleprocessing software runs on the mainframe and handles all traffic between the CPU and the network. Its functions include:

- tracking the connections between terminals and application programs
- buffering (or temporarily storing) transmissions between the network and application programs
- dealing with line failures
- sending output to the front-end processor for transmission to specific devices

Since the most familiar telecommunications access programs run on the IBM 360/370 series, we will discuss three of them to illustrate the range of programs available:

- **The basic telecommunications access method** (BTAM) provides only basic line control functions, and an installation programmer must write the local routines for scheduling and allocating resources to terminals, printers, etc. These programs are written in *assembler language*.
- **The telecommunications access method** (TCAM) can handle data communications in a more complex multiprogram system, and can also manage a *system network architecture* (SNA) network. TCAM has its own control program (written in assembler language) for handling message traffic which does all the handshaking, polling, editing, routing, queuing, and the like.
- **The virtual telecommunications access method** (VTAM) is an advanced communication software package intended to support SNA networks. VTAM employs a higher-level *macro language* which lets the user define the network to meet local needs. VTAM also controls message traffic between application programs and remote terminals, begins and ends sessions between application programs and terminals, and helps application programs share circuits and other resources. VTAM's programming can be altered to accommodate a new network configuration or other changes.

————— PERFORMANCE SOFTWARE ——————————

Those individuals responsible for the smooth operation of the network need to be able to diagnose problems and monitor performance. Fortunately, there are programs that do exactly this. These may be run on the CPU, on the front-end processor, or on special equipment attached to the network, and they usually provide a graphic representation of real-time traffic on the system as well as line performance and error diagnostics (along with automatic error alarms).

————— APPLICATION PROGRAMS ——————————

Application programs are the reason why a company acquires a computer in the first place: they provide the services (e.g., accounting, marketing, production) the computer user wants. The network exists *solely* to provide access to these services.

Application programs are usually the responsibility of computer programmers, who tailor the software to the needs of individual users.

THE BRISFIELD COMPANY

Briefly describe your estimate of the network software needed to run the Brisfield Company's CyberQuote service (see pages 18–19).

SUMMARY

1. Software falls into two major categories: system software, which helps operate and control the computer, and application programs, which solve specific user problems.
2. Between the system software and the application programs there are two additional layers of software: telecommunications access software and teleprocessing monitor software.
3. Telecommunications access software allows application programs access to the network so they can send and receive data.
4. Teleprocessing monitor software is the interface between the telecommunications access software and the application programs.
5. Data bases are collections of data organized to avoid duplication and permit retrieval of information for a wide variety of uses.
6. The software that maintains data bases is called a data base management system.

7. The important elements in networking software include line control, teleprocessing control, and performance monitoring.

8. The network control program uses a network definition data base to operate the network.

REVIEW QUESTIONS

1. Briefly discuss software, protocols, and network architecture.

2. What are operating systems and application programs?

3. Name the three kinds of programs that make up an operating system.

4. Explain job, task, and data management.

5. What are some basic functions of a telecommunications access method?

6. Explain basic telecommunications access method, telecommunications access method, and virtual telecommunications access method.

7. Describe a teleprocessing monitor and IBM's Customer Information Control System.

8. Define data base and data base management system (DBMS).

9. Why are DBMSs important?

10. Describe the four important categories of networking software.

11. What is the function of a network definition data base? What is it composed of?

12. Discuss the functions of teleprocessing software and explain how three standard software programs work.

SPENCER GIFTS TAKES CREDIT A STEP FARTHER
Authorization System Readies Chain for Electronic Data Capture

Spencer Gifts, the Pleasantville, N.J.-based retailer, has taken the first step into the future of credit authorization by subscribing to TeleCard, a credit authorization service of Telenet, a subsidiary of U.S. Sprint. While the chain is currently an authorization-only subscriber, it is only a matter of time before electronic data capture is added to Spencer's transaction processing.

The company, which operates over 500 stores under the names of Spencer Gifts (giftware), A to Z (specialty gadgets à la Brookstone or Sharper Image), and Intrigue (jewelry), has installed the system in 167 locations: 60 Spencer Gifts, seven A to Z, and 96 Intrigue.

Before the installation, the company had been doing credit authorization via manual dial-up wire to the appropriate credit authorization center or by consulting printed listings of bad account numbers. Today, these 167 locations utilize Micro-Fone II credit authorization terminals which access the Telenet data network. The network forwards the consumer account and purchasing information to either Telenet's host computer, a Tandem TXP system, or an intermediary service.

Either the Telenet host or the intermediary service forwards the information to the appropriate authorization data base. If the transaction is authorized, an approval code is transmitted back over the same network to the Micro-Fone terminal.

According to Tom Best, director of market research, the decision to install the system was based on customer service and financial considerations. The customer service benefits came from slashing the time to actually conduct the credit authorization transaction processing down to 15 to 20 seconds. "We wanted to speed the customer through the cash and wrap," he explains.

On the financial side, there were multiple benefits to be realized.

Like many retailers with a manual authorization operation, Spencer Gifts was incurring charge-backs due to human error. "There's going to be human error when going through a book and looking for a number—especially with the size of print those books are published in," he explains. "Occasionally we would realize charge-backs from the respective credit card companies because we did not follow procedure."

Additionally, the three chains do 100% authorization "whether it's a $2 sale or a $300 sale and that's part of our agreement with Telenet. When you do that, you virtually eliminate any charge-backs."

Another financial consideration was based on the hardware. The Micro-Fone II terminals come equipped with telephones. Until the terminals were installed, Spencer Gifts was leasing all of its telephones from the various telephone companies. "When we purchased the Micro-Fones," says Best, "we returned the equipment to the respective phone companies and avoided those monthly lease charges."

Aside from the hardware, the company is also realizing savings on telephone usage because it is no longer paying for its clerks to dial into a voice authorization center manually. While he could not quote specific figures, Best says that the amount was enough to financially justify the installation. Now Spencer Gifts pays 9¢ per call for local node access and 23¢ for 800 number access. He does note, though, that just "on a local basis, there's not much savings."

What all these benefits add up to is a 46% return on investment on the initial purchase, with a breakeven of two years and two months, says Best.

This installation is only the first phase, however. Ultimately Spencer Gifts plans to subscribe to electronic data capture. The Telenet system allows for the host computer to capture the credit data as the approval code is being transmitted back to the retailer's terminal. The host would then electronically transmit the data to the retailer's bank which provides for reimbursement. Conceiv-

From *Chain Store Age Executive*, February 1987, 57. Reprinted with permission.

ably, this could knock five to seven days off the time it takes for reimbursement to be made.

Additionally, notes Best, "when you go to data capture you can use a black box to tie the Micro-Fones into the POS system. This would allow us to realize other accounting and check and balance opportunities—and eventually we will probably get to that."

6 PROTOCOLS

HANDSHAKING
BISYNC
SDLC

AFTER READING THIS CHAPTER, YOU WILL UNDERSTAND THAT:

- Handshaking is the signal exchange between devices that establishes the circuit for transmission.
- Once handshaking establishes the circuit, transmission is subject to line discipline.
- Line discipline controls the way information is transmitted and received (including data block sequencing and error control procedures).
- Two widely used line disciplines are Bisync (binary synchronous communications protocol) and SDLC (synchronous data link control).

The dictionary says that a *protocol* is "a code prescribing strict adherence to correct etiquette and precedence (as in diplomatic exchange and in the military services)." Similarly, **protocols** in telecommunications define how networks establish communications between devices, exchange information, and terminate communication. The three areas covered by protocols are:

1. *Device control.* Personal computers, terminals, printers, and so on, can all be connected to a network, and each device has its own specific capabilities, such as carriage returns, line feeds, and backspaces. The commands necessary to control these features are specified in the *device control protocol* for each device.

2. *Data link control.* Each transmission of information between points in the network is regulated by the *data link control protocol,* which initiates, checks, and (if necessary) retransmits the transmission.

3. *Data format control.* The exact format and content of messages (data exchanges as well as control messages), including headers, text delimiters, and error checking, are specified by the *data format control.*

Protocols are thus the conventions and procedures required to open and maintain communications. In operation, protocols have two major functions:

■ *handshaking,* the signal exchange between devices that establishes a logical communication channel for information transmission
■ *line discipline,* which controls the way information is transmitted and received, including how data blocks are sequenced and errors are controlled.

That is, once handshaking establishes the channel, transmission is subject to the line discipline.

Before discussing handshaking, we need to agree on a couple of concepts. First, from a protocol standpoint, half-duplex is a protocol for transmitting information in one direction at a time between two points, and full-duplex is a protocol for transmitting information in both directions simultaneously between the same two points (see Figure 6–1).

Second, for two network components to operate together effectively, they need to be compatible at various levels, including:

■ *mechanical* compatibility, including matching plugs, sockets, and cables
■ *electrical* compatibility, so the right wires carry the right signals at the right time
■ *hardware* that can put information on the channel at one end and take it off at the other

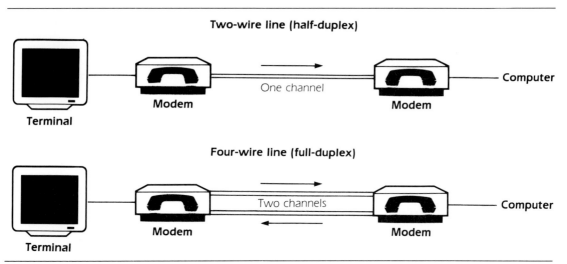

FIGURE 6–1 *Half-duplex and full-duplex.*

▪ *software* that integrates the programs running on all the devices.

Achieving this kind of compatibility at all such levels is one of the major considerations in business telecommunications, as we will see in Chapter 7.

——————— HANDSHAKING ———————

Handshaking occurs prior to the actual data transmission, and its purpose is to ensure that an operational circuit is available. There are different kinds of handshaking for different circuits, and we will concentrate on the handshaking used by modems over two-wire lines and four-wire lines (see Figure 6–1).

A two-wire line is usually a single channel and half-duplex, permitting alternate, one-way-at-a-time communication. That is, a terminal sends a message down the line to the central computer, which receives the message, processes it, and transmits a response back to the terminal using the same channel. Since there is only one communication channel between the two modems, only one modem can transmit at a time.

The handshaking sequence for a standard RS 232C interface (see Chapter 2) is as follows:

▪ a message is entered at a terminal (which stores it in its internal buffer) and the transmit key is pressed when it is ready to send

- the terminal sends a *request-to-send* signal to the transmitting modem, which tells it that the terminal wants to transmit (see pin number four in Figure 2–11)
- the transmitting modem sends a carrier wave down the telephone line
- the receiving modem locks onto the carrier wave, synchronizes itself with the signal, and sends the central computer a *data-carrier-detect* signal (pin number eight in Figure 2–11), telling the computer it is ready to receive data
- the transmitting modem sends a *clear-to-send* signal (pin number five) to tell the terminal to start sending
- the terminal transmits its block of data, which the transmitting modem modulates onto the carrier wave
- the receiving modem demodulates the data and passes it on to the computer, which validates the data
- when the terminal is finished transmitting, it stops sending the request-to-send signal, which causes the transmitting modem to stop sending the carrier wave and the clear-to-send signal
- the receiving modem then stops sending the data-carrier-detect signal to the computer.

When the computer has a response for the terminal, the same process is repeated in reverse.

Note that handshaking only establishes the circuit; the actual data transmission is controlled by the protocol (e.g., Bisync or SDLC, discussed below).

A four-wire line has two channels and is full-duplex, allowing transmission in two directions simultaneously. Full-duplex modems can send their carrier signals continuously; since they are on different channels they do not interfere with each other. Therefore, when the terminal wants to transmit, it can send the signal straight down the line to the computer. Full-duplex transmission is much more efficient than half-duplex, and avoids many of the delays inherent in a half-duplex network.

BISYNC

For telecommunications to occur, the communicating devices have to conform to the same conventions or protocols, conventions which include data coding, error handling, and timing and sequencing. Bisync (binary synchronous communications protocol) is such a protocol that has been widely used since 1968 for transmission between IBM computers and data communication hardware. Bisync supports both ASCII and EBCDIC codes.

Bisync is a line discipline—a set of rules specifying how information is exchanged on the network—that uses standard control characters and sequences for the synchronous transmission of binary-coded data between computers or between terminals and computers. It is byte- or character-oriented, and uses the entire eight-bit byte to send a command signal to the receiving station (in contrast with newer bit-oriented protocols such as SDLC).

Using the Bisync format, a message is surrounded by a header and trailer as shown in Figure 6–2. A typical Bisync data exchange might occur as follows (note the half-duplex movement, like a ping pong ball):

TERMINAL	COMPUTER
1. terminal indicates it has data to send by transmitting an ENQ (enquiry) control character	
	2. computer receives the ENQ control character
	3. computer responds with an ACK (acknowledgment) control character, which tells the terminal to transmit
4. terminal receives the ACK control characters	
5. terminal transmits a block of data	
	6. computer receives the block and checks for errors. If there are no errors, go to step 8; if an error is found, computer sends a NAK (negative acknowledgment) control character
7. terminal receives the NAK and retransmits the last block	
	8. computer acknowledges receipt of correct block with an ACK, which also tells the terminal to send the block
9. terminal sends the next block. If message is finished, it sends an EOT (end-of-text) control character	
	10. computer receives the EOT and stops receiving.

The ASCII and EBCDIC code tables (Figures 2–7 and 2–9) show the special control characters used in Bisync. The acknowledgment control characters alternate between ACK0 (for even-numbered blocks) and ACK1 (for odd-numbered blocks). When the acknowledgment and block numbers do not match, the blocks are retransmitted (this helps ensure that no block or acknowledgment is lost, as might happen because of power surges on the line).

FIGURE 6-2

Bisync message format.

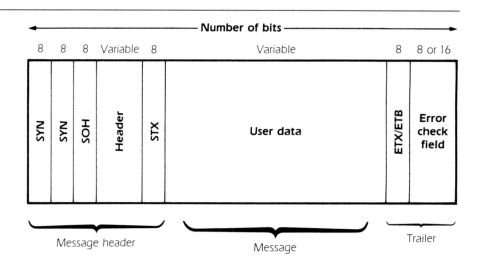

SYN Synchronization characters ensure that the transmitter and receiver interpret the data stream identically.

SOH Start of header (optional).

HEADER User-provided information (e.g., terminal address, author, title of message).

STX Start of text (i.e., beginning of message).

USER DATA The basic information being communicated.

ETX End of text, or end of user data.

ETB End of block: when user data is divided into multiple blocks, ETB is sent at the end of each block, except the last one.

ERROR CHECK ASCII error checking uses a Vertical Redundancy Check (VRC) and a Longitudinal Redundancy Check (LRC); EBCDIC uses Cyclic Redundancy Checking (CRC).

Bisync is used for low-speed transmissions of, say, up to 2400 bits per second, and almost always over dial-up circuits. Bisync cannot take advantage of full-duplex lines.

SDLC

SDLC (synchronous data link control)* was introduced by IBM in 1973 for communication to bank teller terminals. Since two or more terminals may be connected to one line (see Figure 6–3), the computer may have to transmit data to one terminal while simultaneously receiving data from another terminal on the same line. SDLC, an offshoot of the international

*A data link includes the physical medium of transmission, the protocol, and associated devices and programs.

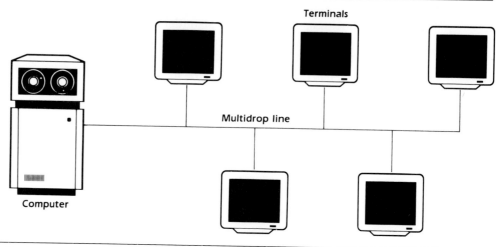

FIGURE 6-3 *Multidrop line.*

standard HDLC (high level data link control) manages information transfer over such data communication links.

SDLC messages are sent across the line in a specified format called a *frame*. The line procedures are based on a pure binary data stream, so SDLC is not dependent on any particular code such as ASCII or EBCDIC (although it requires that all fields in the frame be sent in multiples of eight bits). All the bits are precisely timed, and the start of the message is flagged by a unique bit pattern, the *flag bit* of the frame.

SDLC defines the line discipline for transmitting and receiving information, but does not define the handshaking sequence which establishes the transmission circuit. Handshaking must occur before the transmission of the first flag, and only then can a computer and a terminal begin to exchange data according to the SDLC protocol.

SDLC is implemented by printed circuit boards, firmware, and software at each end, and is transparent to the user. The flag and other bits which are put before and after each message at the transmitting end are removed at the receiving end, so only the original bit stream—the message—is passed on to the user.

The format of a frame is shown in Figure 6-4.

Flag 01111110	Address	Control	Message	Frame check sequence	Flag 01111110

FIGURE 6-4 Frame format.

- The *flag bit* is both the first and last byte. It is always the binary sequence 01111110, and if a sequence of six ones occurs elsewhere in the transmission, the transmitter stuffs a zero after the fifth one to prevent the receiver from mistaking these bits for a flag (the receiver then removes this extra zero).
- The next eight-bit byte is the *address* of the receiving terminal on the communications line, or link. Frames sent by terminals can be addressed only to the central computer.
- The *control field* is also an eight-bit byte which defines the function of the frame (see below).
- The *message field*—the actual data—varies in length, but is always sent in multiples of eight bits.
- The *frame check sequence* is a 16-bit calculation that is computed by the transmitter from the contents of the address, control, and information fields. This calculation is called a *cyclic redundancy check*, and the receiver does its own calculation and compares it to what was received. If they do not match, that means there were transmission errors, and retransmission is requested (see Chapter 12).

SDLC frames come in three types: information frames (which transfer information), supervisory frames (which indicate that the line is busy or ready, report frame numbering errors, and confirm that data is being received), and nonsequenced (unnumbered) frames (which initialize and disconnect stations). Only information frames can be used to transfer data; the other two types help control the flow of information along the link. The control field, as noted, defines the function of each frame. The first and second bits identify the frame type (e.g., a zero in the first bit position means that the frame is an information frame). The three control field formats are described in more detail in the next sections.

Information transfer format

Figure 6–5 illustrates the information transfer format. Ns is the sending device's sequence number for the block being sent. When transmitted by the central computer, P/F is called the P bit (poll message bit). When set at one, the terminal should answer, and when set at zero, the terminal should not reply. When transmitted by a terminal, P/F is called the F bit

FIGURE 6–5

Information transfer format.

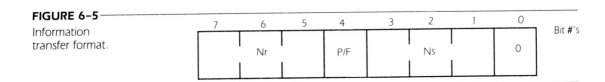

(final message bit). When set at zero, there is at least one more block to transmit, and when set at one, this block is the last of the current sequence to be transmitted by that terminal. The P/F bit helps the computer to send and receive data from different terminals on the same line simultaneously (see Figure 6–3). For instance, a computer is sending a multiblock transmission to Terminal 1 while receiving a multiblock transmission from Terminal 2. It receives an F bit from Terminal 2 when the current block has been completely transmitted, so the computer interrupts its transmission to Terminal 1 long enough to poll Terminal 3 for whatever it has to send. When Terminal 3 begins sending, the CPU resumes transmission to Terminal 1. Nr is the number of the next block which the sending device expects to receive from the other end when it next transmits. This, in effect, acknowledges correct receipt of all frames sent so far.

Supervisory format

Figure 6–6 illustrates the supervisory format. *Mode* signals whether a particular terminal is ready to receive or not.

FIGURE 6–6

Supervisory format.

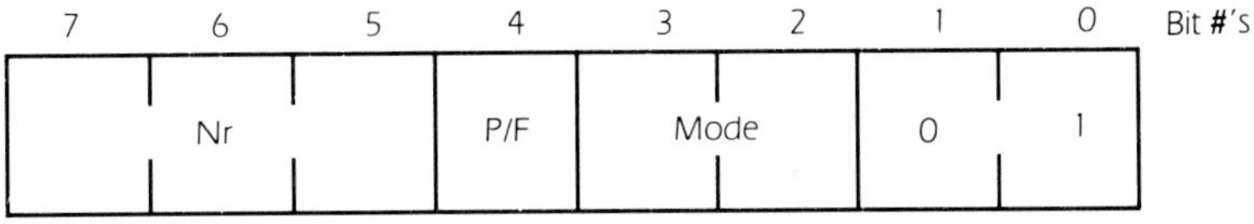

Nonsequenced (unnumbered) format

Figure 6–7 illustrates the nonsequenced (unnumbered) format. Bits 2, 3, 5, 6, and 7 are taken as a unit, and when transmitted by the central computer, these C/R bits are used to send a command to a terminal (for instance, to achieve initial synchronization or to disconnect). When sent by a terminal, the C/R bits are used for a response to the computer (for instance, a request to go on-line).

The sequence of frames between sender and receiver is managed by the Ns and Nr bits of the control field, which are set to zero at the start

FIGURE 6–7

Nonsequenced format.

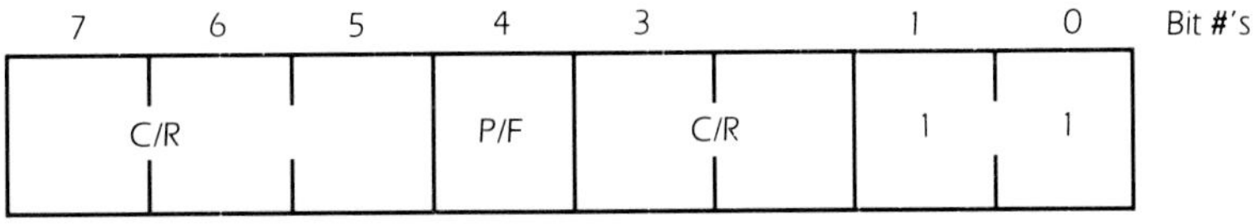

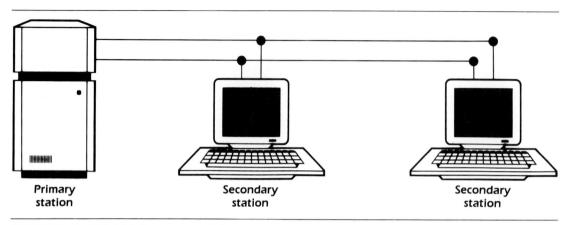

| Primary | Secondary | Secondary |
| station | station | station |

FIGURE 6–8 *SDLC Multipoint transmission.*

Adapted from **Synchronous Data Link Control** (IBM), Manual No. GA27–3093–2. See also Ken Sherman, **Data Communications: A User's Guide,** 2nd ed. © 1985 Prentice-Hall, Inc., Englewood Cliffs, New Jersey.

of each transmission between the central computer and a terminal. During the subsequent transfer of frames, the sender counts the outgoing frames and transmits this count via the Ns bits of the control field. The receiving station then checks the number of frames. If the count is accurate, the receiver increases its Nr count by one and returns an acknowledgment containing this new Nr count to the sender. At this point in the exchange, the receiving station's Nr field (the next frame it expects to receive) should be one larger than the transmitting station's Ns field (the number of frames transmitted so far).

An actual transmission with the SDLC protocol is illustrated in Figure 6–8 and described in the Appendix to this chapter. The article at the end of this chapter, "Ruling on X.25 Puts BOCs in Key Spot to Handle Corporate Data," illustrates that there is much more to protocols (e.g., routing, flow control, session establishment, and internetworking) than an introductory chapter can discuss. X.25 (as discussed in Chapter 4) is the CCITT protocol for terminals operating in the packet mode on public data networks.

SUMMARY

1. Telecommunications protocols determine how networks establish communications between devices, exchange information, and terminate communications.

2. In operation, protocols have two major functions: handshaking (establishing the circuit for transmission), and line discipline (the ac-

tual transmission, including data block sequencing and error control procedures).

3. Two widely used line discipline protocols are Bisync (binary synchronous communications protocol) and SDLC (synchronous data link control).

4. Bisync is a half-duplex protocol that transmits strings of characters in one direction at a time at low speeds over dial-up circuits.

5. Bisync messages each have a header and a trailer. Each transmission is answered by an acknowledgment.

6. SDLC is a more efficient full-duplex protocol that can transmit in two directions simultaneously.

7. SDLC messages are sent in frames, and the bits put before and after each message by the transmitter are removed at the receiving end, so only the message is passed on to the user.

REVIEW QUESTIONS

1. Define protocol, handshaking, and line discipline.
2. Describe the handshaking sequence for a RS 232C interface, for Bisync, and for SDLC.
3. Describe a Bisync exchange.
4. Why does Bisync use ACK0 and ACK1 control characters?
5. Briefly, how does SDLC differ from Bisync?
6. What is a frame?
7. What are the control field formats for SDLC?
8. What is the function of the Ns and Nr bits?
9. Explain the SDLC sequence in the Appendix.
10. The article beginning on page 106 talks about the X.25 protocol. What is it? Why is it important?

ENDNOTE

1. *Webster's New Collegiate Dictionary* (1981).

APPENDIX
SDLC Transmission

An actual transmission with the SDLC protocol is illustrated in Figure 6–8. We assume the following synchronization sequence has already occurred:

1. The central computer (A) polls the terminal (C) ("Do you have anything to send?").
2. C requests a session ("Yes").
3. A sets C to normal response mode (i.e., A controls the flow and C cannot send unsolicited frames).
4. C acknowledges.

We will use these symbols to represent the frames:

F, A, C, Ns, P/F, Nr, FCS, F

Where: F = Flags
   A = Address of secondary station
   C = Command/Response Acronym
   Ns = Count of sending sequence field
   P/F = P: poll bit on
      $\overline{P}$: poll bit off
      F: final bit on
      $\overline{F}$: final bit off
      Nr: Count of receiving sequence field
      FCS: Frame check sequence field

1. A polls C (the computer tells terminal C it is ready to receive)

   F, C, RR, −, P, 0, FCS, F

   **F: flag
   C: address (terminal C)
   RR: computer is ready to receive
   −: field not used in this type of frame
   P: computer is polling terminal C
   0: the computer has received no frames from terminal C this sequence (Nr = 0)
   **FCS: frame check sequence field
   **F: flag

2. C sends frames 0 through 2 to A (terminal C sends 3 frames to the computer)

   F, C, I, 0, F, 0, $\overline{F}$CS, F

C: address (terminal C)
I: this is an information frame
0: this is the first frame sent (Ns = 0)
$\overline{F}$: there is at least one more frame
0: terminal C has received no frames from the computer (Nr = 0)

   F, C, I, 1, F, 0, FCS, F

All identical to the above except that this is the second frame sent, so Ns = 1

   F, C, I, 2, F, 0, FCS, F

Ns = 2 and since the F bit is sent, this is the final message of the sequence

3. A sends frames 0–1 to B (while step 2, above, is occurring, the computer sends two frames to terminal B)

   F, B, I, 0, P, 0, FCS, F

B: address (terminal B)
I: this is an information frame
0: this is the first frame sent (Ns = 0)
$\overline{P}$: terminal B should not reply
0: the computer has received no frames from terminal B this sequence

   F, B, I, 1, P, 0, FCS, F

As above, except this is the second frame sent (Ns = 1)

4. A asks B for acknowledgment (the computer asks: "terminal B, did you get those 2 frames OK?")

   F, B, RR, −, P, 0, FCS, F

B: address (terminal B)
RR: computer is ready to receive an acknowledgment
−: field not used in this type of frame
P: computer expects a reply from terminal B
0: the computer has received no frames from terminal B this sequence

**These remain the same throughout, so we do not repeat these explanations.

5. B ACK's frames 0–1 (terminal B tells the computer that frames 0 and 1 were received correctly)

F, B, RR, −, F, 2, FCS, F

B: address (terminal B)
RR: terminal B is ready to receive
−: field not used in this type of frame
F: this is the final message of this sequence
2: terminal B has received frames 0 and 1 correctly from the computer

6. A ACK's frames 0–2 (the computer tells terminal C that frames 0, 1, and 2 were received correctly)

F, C, RR, −, P, 3, FCS, F

C: address (terminal C)
RR: the computer is ready to receive
−: field not used in this type of frame
$\overline{P}$: terminal C is not expected to reply
3: the computer has received frames 0, 1, and 2 correctly from terminal C

7. A sends frame 0 to C (the computer starts another transmission sequence to terminal C)

F, C, I, 0, P, 3, FCS, F

B: address (terminal B)
I: information frame
2: this is the third frame sent this sequence
$\overline{P}$: Terminal B is not expected to reply
0: the computer has received no frames from terminal B this sequence

9. A sends frame 3 to B and asks for acknowledgment (the computer sends the fourth frame in this sequence and asks terminal B to acknowledge)

F, B, I, 3, P, 0, FCS, F

B: address (terminal B)
I: information frame

3: this is the fourth frame sent to B this sequence
P: terminal B is expected to reply
0: the computer has received no frames from terminal B this sequence

10. B ACK's frames 2–3 (terminal B acknowledges correct receipt of frames 2 and 3)

F, B, RR, −, F, 4, FCS, F

B: address (terminal B)
RR: terminal B is ready to receive
−: field not used in this type of frame
F: this is the final message of this sequence
4: terminal B has correctly received frames 0, 1, 2, and 3 in this sequence

11. A sends frame 1 to C and asks for an acknowledgment (the computer sends the second frame in this sequence to terminal C and asks if frames 0 and 1 were correctly received)

F, C, I, 1, P, 3, FCS, F

C: address (terminal C)
I: information frame
1: this is the second frame sent to C this sequence
P: terminal C is expected to reply
3: the computer has received frames 0, 1, and 2 correctly from terminal C this sequence

12. C ACK's frame 1 (terminal C acknowledges that frames 0 and 1 were received correctly from the computer this sequence)

F, C, RR, −, F, 2, FCS, F

C: address (terminal C)
RR: terminal C is ready to receive
−: field not used in this type of frame
F: this is the final message this sequence
2: terminal C has correctly received frames 0 and 1 in this sequence

This section draws heavily on *Synchronous Data Link Control* (IBM), Manual No. GA27–3093–2.

RULING ON X.25 PUTS BOCs IN KEY SPOT TO HANDLE CORPORATE DATA

In March the Federal Communications Commission (FCC) conditionally ruled that seven Bell Operating Companies (BOCs) could perform asynchronous/X.25 data protocol conversion as one of their basic telecommunications offerings. The ruling means the BOCs will play a greater role in handling corporate data traffic and may solve the "last mile" problem that often troubles end users. Some communications consultants predict that both value-added network companies (VANs) and modem manufacturers will face a powerful new competitor.

Under this FCC waiver, the BOCs are free to convert data using the asynchronous/X.25 protocol as part of regular operations. The Bell companies involved are Pacific Bell, Southern and South Central Bell, Southwest, New York and New England Telephone, New Jersey and Pacific Northwest Bell, and Ameritech. The FCC decision modified an earlier ruling, known as Computer II, which required the BOCs to set up separate and distinct subsidiaries if they wanted to provide data conversion and packet-switching networks. The BOCs had argued that unless they could offer these data services as part of basic operations the cost to potentially new customers—such as home users—would be too high.

The X.25 protocol continues to be the most widely used international packet-switching protocol that is compatible with IBM's system network architecture (SNA). Packet switching has many advantages, including lower costs for data transfer and increased network reliability because data from different computers can be bundled together.

"The VANs made a number of contentions and didn't want to see the waiver granted," said Mike Slomin of the FCC's Common Carrier Bureau. "The record is clear that the BOCs will serve local needs and the VANs are in the long-distance market." But Slomin added that the FCC recognized that in granting the waiver, an edge might be given to the BOCs that local phone companies hadn't had previously.

Janis Langley of Bell Atlantic's management services division expects her company to offer large- and medium-sized business customers rate reductions while expanding data communications services, including access to a range of data bases. "A lot of companies with interactive communications will see dramatically reduced costs," Langley predicted. "It will also be attractive for small- and medium-sized companies because they won't have to purchase expensive interface devices—such as modems—to use the local network."

Telecommunications analyst Steve Caswell of Trigon systems in Canada suggested that local phone companies may build in their own modems, which carry both voice and data, and automatically send the data to the packet network. "You may never need a modem," Caswell said, "but this development will take place over many years." Caswell added that the FCC's decision could keep the VANs locked out of the lucrative local data market.

Dan Rosenbaum of the Yankee Group, a consulting firm in Boston, believes that both VANs and BOCs will benefit from the competition fostered by this decision. Because the BOCs can't offer long-distance services, the VANs will continue to provide data transmission. "The BOCs also can't provide information [like a VAN]. There's far more offered by a VAN to a customer than just protocol conversion."

The BOCs hope that residential consumers will benefit from their involvement in data transmission and will make the home information market a reality. They predict that the FCC ruling will allow them to cut consumer computer-to-computer costs in half. This could spur dramatic growth in consumer data communications services such as home banking.

The BOCs expect to continue their push as net-

Reprinted from *Infosystems*, May 1985, 16. Copyright Hitchcock Publishing Company.

work providers and innovators. At the end of March, Pacific Bell announced it had applied for a patent on a product that converts a single phone line into two voice and data channels, thus providing simultaneous voice, data, and telemetry transmissions. This continued integration of computers and telephones only seems to stimulate the appetites of AT&T's children. Bell Atlantic's Langley added: "The local data service is as potentially revolutionary as the electronic switch was to plain old telephone service. We're staking out a whole new terrain of the information age."

7 NETWORK ARCHITECTURE

THE OSI MODEL
SYSTEMS NETWORK ARCHITECTURE

AFTER READING THIS CHAPTER, YOU WILL UNDERSTAND THAT:

- Network architecture is a hierarchical structure, each layer providing a set of functions that can be accessed and used by the layer above it.
- This logical structure of formats, protocols, and operational sequences is transparent to users.
- Each layer is independent, allowing a given layer to be changed without affecting other layers.
- The International Standards Organization's seven-layer OSI model defines all the functions needed for two machines to communicate, and has been widely accepted by the telecommunications industry.
- IBM's Systems Network Architecture provides the same functions as the ISO model, but implements the functions somewhat differently.

Network architecture encompasses all the parts of a network, including hardware, software, protocols, topologies, and access methods, as well as how these parts communicate and cooperate, and how they fit together to form the network. As shown in Figure 7–1, a telecommunications network cannot operate without the interaction of protocols, software, and architecture: the software must be written to meet both the network standards and the protocol requirements.

As office and plant automation efforts grew in size and complexity and spread out geographically, communications hardware and software had to handle more functions. But as new tasks were grafted onto already overloaded systems, communications often became very slow and cumbersome. The solution was to divide and conquer: reduce the complexity by grouping similar functions into self-contained layers, and allowing interaction only between adjacent layers.

This hierarchal structure with different layers for various functions and services tends to move telecommunications away from incompatible, one-time-only solutions and toward generic solutions, independent of any particular hardware or software products. Since each layer is relatively self-contained, with clearly defined interfaces, a change made in one layer need not affect other layers. The OSI model, which we will discuss next, enables machines from different vendors to communicate even if they are on different continents.

FIGURE 7–1

Relationship of
protocols, software,
and network
architecture.

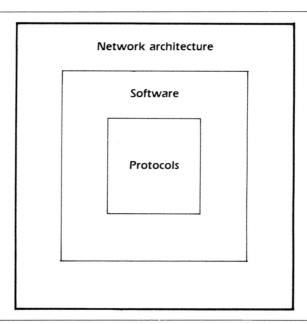

THE OSI MODEL ———————————————————

Communications between equipment made by different manufacturers (and sometimes between different models made by the same vendor) are complicated by incompatible data formats, data exchange conventions, and other such differences. A unique hardware and software solution for each situation would be costly; a much better alternative is to agree on a common set of conventions.

The national standards organizations of many countries, including the American National Standards Institute, belong to the International Standards Organization (ISO). In 1977, the ISO began work on an *open system interconnection* (OSI) model to deal with these incompatibilities. Finally, in July 1979, the seven-layer OSI standard was announced. Since then, it has been widely accepted by the telecommunications industry.

This OSI model (see Figure 7–2) takes into account hardware, protocols, software, and network architecture, and although the actual connections between the computer, terminals, and other devices are physical, the connections between layers in the model are logical, not physical. Together, these seven layers define the functions needed for any two machines to communicate.

Layers 1 and 2 depend on both hardware and software, while the upper layers depend entirely on software. The software used for layers 4 through 7 generally resides on the mainframe, and the software for layers 2–4 usually resides on the front-end processor.

Communications between the mainframe and a terminal must pass through all seven layers at each end. A terminal user who is accessing a data base, for example, would get the information as illustrated in Figure 7–3. The mainframe's data base software passes the data to the application layer, which adds an *AH* header. This header tells the application layer at the terminal end what to do to perform its functions. The data plus AH header are then passed as a unit to the presentation layer, which adds its own PH header. This continues down through the data link layer, which usually adds both a header and a trailer to create a frame, similar to the SDLC frame on page 99. This frame is then transmitted as a series of bits. When the frame is received at the terminal end, the reverse process occurs: each layer removes its header and acts on the instructions therein, passing the remaining data up to the next level.

In order for hardware and software from different vendors to communicate, they have to use these same seven layers, and the data formats and control fields passing between the layers have to be compatible.[1]

Let's look at each of the seven layers in detail.

Physical control

The function of layer 1, the physical layer, is to pass bits (zeros and ones) from one device to another, initiating, maintaining, and terminating the

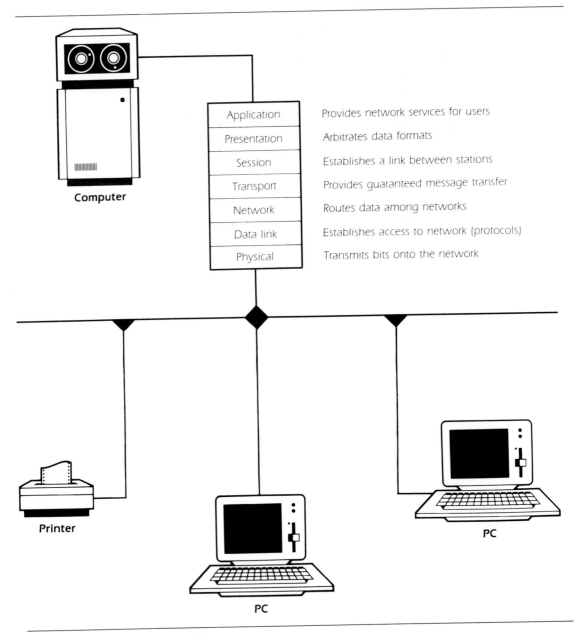

FIGURE 7–2 OSI reference model.

connection. The connection is physical only at this layer; there is no actual connection between, say, level 2 of the mainframe and level 2 of the terminal. Layer 1 is concerned with voltages, timing (transmission speed in bps), full-duplex or half-duplex transmission, connector cable standards (such as RS-232C), and the other issues discussed in Chapter 2.

Data link control

Layer 2, the data link layer, manages the basic connection created by layer 1, transforming it into a circuit or link. The raw bit stream is broken into data frames, which are transmitted, received, and acknowledged. This layer also provides error detection and control, assuring layer 3 of error-free transmission. Since layer 2 sends blocks of data over the physical link of layer 1, it is concerned with the line control procedures discussed in Chapter 6.

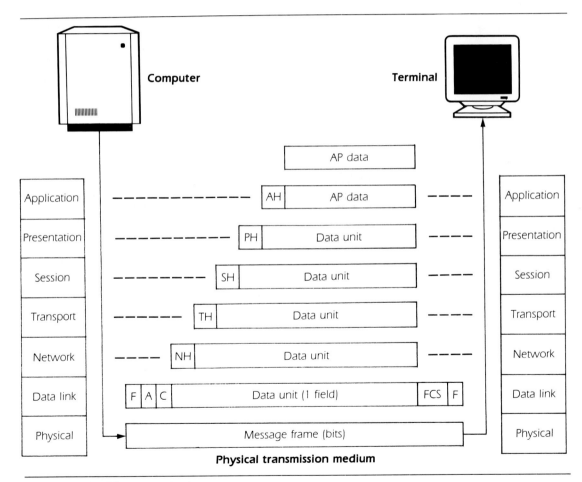

FIGURE 7–3 OSI communication.

Adapted from **A Primer on Computer Networks** (Wang, 1982), Publication No. 777–4004, 1st edition, 61. Courtesy Wang Laboratories, Inc.

Network control

Layer 3 controls the combined operations of the first three layers, and supplies such services as traffic control and routing. The network layer is also responsible for establishing, maintaining, and terminating connections between the various devices, switching and routing the data as necessary. Essentially, layer 3 deals with the details of moving data through the network, which leaves the upper layers independent of the transmission and switching technologies used to connect devices.

Transport control

Layers 1, 2, and 3 operate in a chained manner, transferring data from terminal to terminal (node to node) in a methodical fashion, never skipping a terminal or sending data over a different route. In contrast, layer 4 is concerned with the transmission of data from source to destination, and ensures that the data is delivered error-free and in sequence. In effect, the transport layer allows a program on the mainframe to converse directly with software running on the destination machine, even though they are separated by modems, multiplexers, front-end processors, and other hardware. Layer 4 does not deal with the details of the underlying network; it is concerned only with end-to-end flow control, error recovery, and such user parameters as priority and security. Layer 4 functions are usually carried out by the mainframe software.

Session control

Layer 5, the session layer, controls how two application programs talk to each other: whether the connection is simplex, half-duplex, or full-duplex, and how error recovery is implemented. It is similar to layer 4 in that it operates from source to destination, but is more application-oriented. Layer 5 is normally handled by the mainframe's operating system.

Presentation control

A stream of bits reaching a terminal contains characters which cause tabbing, line feeds, data editing, and the like. Since different devices may have different character sets, layer 6 does the conversion so the devices can communicate.

Application control

The user accesses the network through layer 7, the application layer. It is up to each user and software designer to determine how their programs will plug into the OSI environment for such tasks as data base distribution, initiation and termination of remote devices, file transfer, printing (on a variety of devices), and network statistics. Layer 7 thus relates to specific user applications, and how these applications access the network. All the lower layers exist solely to support layer 7.

Next, we will discuss a commercial implementation of the ISO model: IBM's Systems Network Architecture.

SYSTEMS NETWORK ARCHITECTURE

IBM introduced *Systems Network Architecture*[2] (SNA) in 1974, and it was followed by many others, including Xerox Network Systems and DEC's Digital Network Architecture. SNA is one concept of how data communications should be implemented: it defines the rules for the interaction of computers, terminals, software, and other network components, and allows multiple terminals, for example, to share the files and the processing power of a central mainframe. SNA permits an application program to communicate with any input/output device, and allows application programs and communication equipment to be added or modified without affecting the network. The functions provided to the end user are similar to those provided by the OSI model, but the functions are implemented somewhat differently.

An SNA network (see Figure 7–4) consists of the following:

- a mainframe or central computer
- front-end processors which relieve the mainframe of communications functions such as line control, message handling, code conversion, and error control
- remote controllers, which manage the details of line control and the routing of data through the network
- various general-purpose terminals

SNA is organized around the concept of a *domain*, which means all the resources controlled by a particular *system services control point* (SSCP). The SSCP is the mainframe, which manages the domain and also contains a program called a *VTAM* that controls communications between terminals and the mainframe (see Figure 7–5). The SSCP, together with the end user, is called the *host node*. The *communications controller node* has a front-end processor (e.g., an IBM 3705) with a network control program (see Chapter 4), and the *cluster controller node* has a terminal controller to control peripherals. *Terminal nodes* are individual terminals.

The end user talks to the network through a *logical unit* (LU), as illustrated in Figure 7–6. Logical units are software or firmware instructions which provide network access, and can be found in the terminal, controller, front-end processor, or mainframe. For one end user to talk to another, their logical units have to be connected, and a conversation between two logical units is called a *session*. Each LU has its own name and address.

A session begins when, for example, a terminal operator makes a request to access a data base through a data base management system. The

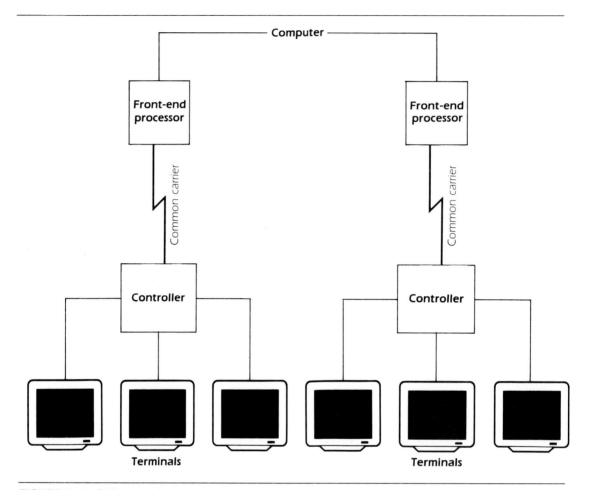

FIGURE 7-4 *SNA network.*

logical units—the terminal and the DBMS—exchange data via frames controlled by the SDLC protocol. The session ends when one of the LUs disconnects.

SNA has two general classes of logical components: *network addressable units* and the *path control network:*

Network addressable units (NAUs) are components such as front-end processors, controllers, and terminals which use software as well as hardware to exchange data and control the network. NAUs communicate through the path control network (see below). An SNA network has three kinds of network addressable units: systems services controller points, logical units, and *physical units* (PUs). Each terminal, controller, or front-

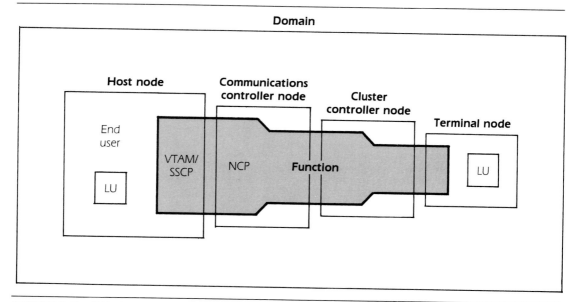

FIGURE 7-5 SNA domain.

From **SNA for Managers** (IBM) Manual No. SRE20-4517-0. Courtesy of International Business Machines Corporation.

end processor is an SNA *node*, and each node contains a physical unit that represents that node and its resources to the system services control point. Accordingly, when the SSCP opens a session with a PU, that node then becomes an active part of the SNA network (see Figure 7-7).

The path control network provides routing (so sessions don't interfere with each other), transmission priorities, flow control (so fast transmitters can't inundate slow receivers), and error recovery. This logical component has two layers:

1. *path control,* which handles routing and flow control
2. *data link control,* which manages data transmission over individual links (via the SDLC protocol).

SNA employs a seven-layer structure similar to the one used by the ISO model. These layers provide the two broad categories of service we noted earlier: network addressable unit services and path control network services (see Figure 7-8). Once again, let's look at each layer.

Physical services

The physical services layer provides the raw bit stream, as in the ISO model.

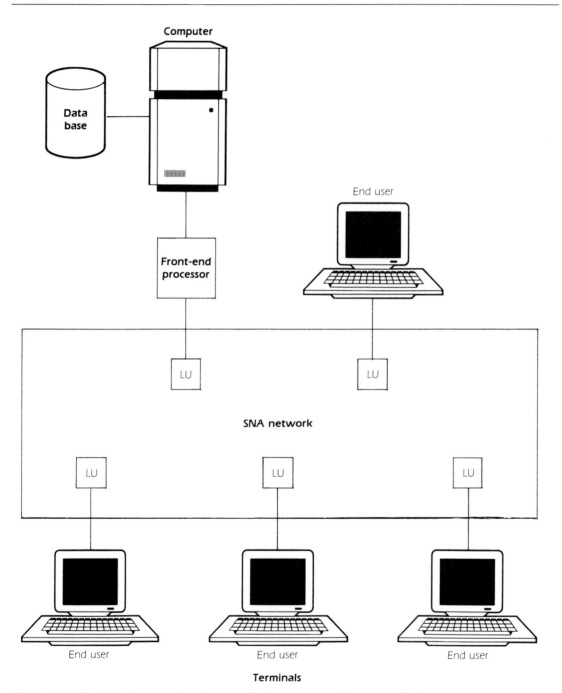

LU: Logical Unit, a bridge between the end user application program and the communications system.

FIGURE 7–6 SNA IU's.

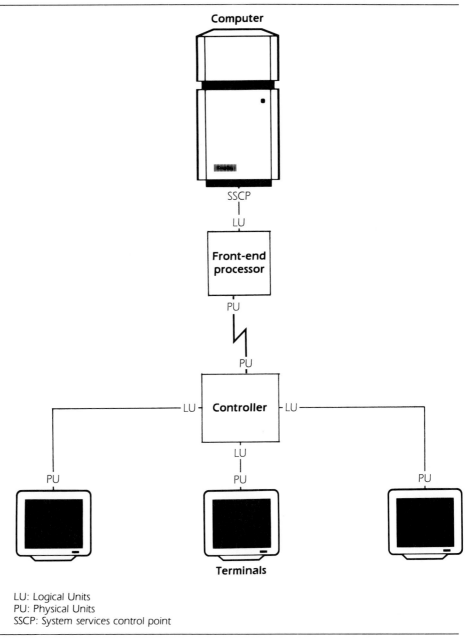

Computer

SSCP

LU

Front-end processor

PU

PU

LU — **Controller** — LU

LU

PU

PU

PU

Terminals

LU: Logical Units
PU: Physical Units
SSCP: System services control point

FIGURE 7–7   *Network addressable units.*

Data link control

This layer breaks up the raw bit stream into data frames according to protocol (see Chapter 6).

FIGURE 7-8
SNA layers.

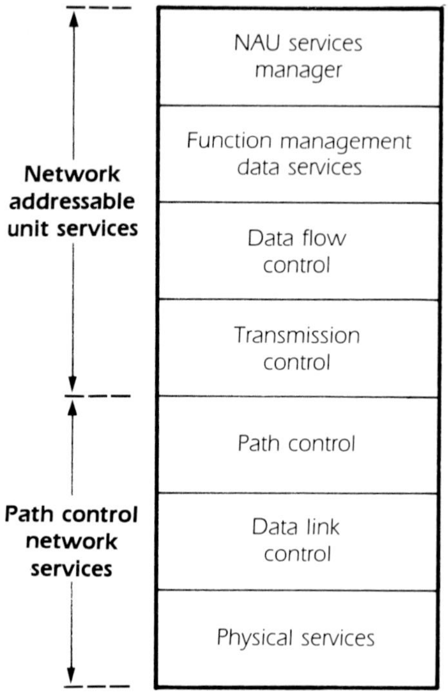

```
        ┌─────────────────────┐
        │    NAU services     │
        │      manager        │
        ├─────────────────────┤
  ↑     │ Function management  │
Network │    data services     │
addressable ├─────────────────┤
unit services │   Data flow    │
        │     control         │
        ├─────────────────────┤
        │   Transmission       │
        │     control         │
        ├─────────────────────┤
        │    Path control      │
        ├─────────────────────┤
Path control │  Data link       │
network │      control         │
services ├─────────────────────┤
        │  Physical services   │
  ↓     └─────────────────────┘
```

Path control

The path control layer creates logical or *explicit* channels between network addressable units (see Figure 7-9). The apparent or *virtual* route between the NAUs looks like a single physical link to higher layers. The path control layer also limits the flow of data to prevent congestion between nodes.

Transmission control

The transmission control layer establishes, maintains, and terminates SNA sessions. It paces the flow of data from source to destination, provides for the sequencing of data blocks, and helps to manage the transmission aspects of a session.

Data flow control

The data flow control layer provides services to terminals, such as specifying the transmission mode, monitoring transmission sequences for recovery, and defining response options.

Function management data services

The function management data services layer provides the end user with *session presentation services*, which translate formats for various terminals

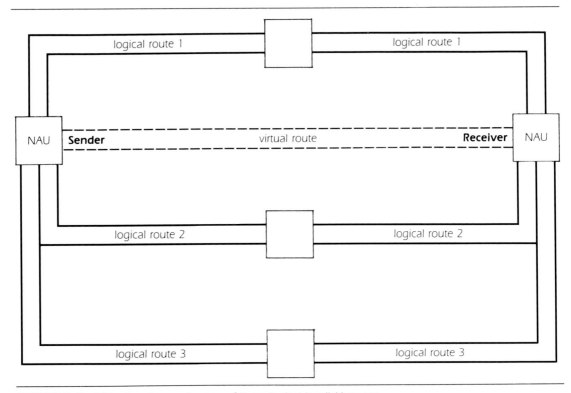

FIGURE 7-9 Virtual route may be any of three logical (explicit) routes.

and compress data to reduce transmission volume; and *session network services,* which configure the appropriate network for a given user, collect and display network statistics, and help diagnose network problems.

NAU services manager

The NAU services manager layer shares the functions of the function management data services layer. A seventh layer is included so there will be a layer-by-layer correspondence between the OSI model and SNA.

Even though the functions provided to the end user are similar, "Mating the Layers of OSI and SNA Will Be Big Job," as discussed in the article at the end of this chapter.

THE BRISFIELD COMPANY

Suggest a workable, less costly approach for the Brisfield Company's Cyber-Quote services (see the case study at the end of Chapter 1).

──────────────────────────────── **SUMMARY**

1. Network architecture is composed of layers of functions.
2. Each layer is independent, so it can be modified without affecting the other layers.
3. End users are unaffected by the layers, formats, and protocols needed to transmit data through the network.
4. The International Standards Organization has developed a seven-layer OSI model for network architecture that has been widely accepted.
5. IBM's Systems Network Architecture provides similar functions to the end user, but the implementation differs.

──────────────────────────── **REVIEW QUESTIONS**

1. Define network architecture, ISO, SNA.
2. Explain Figures 7–2 and 7–8.
3. What is the importance of a standardized network architecture?
4. SNA is organized around domains. Explain.
5. Describe SNA's two general classes of logical components.
6. After reading the article beginning on page 123, comment on IBM's LU 6.2. Why is it important?
7. How are OSI and X.25 related? How are SNA and X.25 related?
8. "OSI in no way ensures that networks will work together." Comment.
9. Why will mating OSI and SNA be a big job?

ENDNOTES

1. See William Stallings, *Data and Computer Communications* (Macmillan, 1985), 388.
2. See *Systems Network Architecture: Concepts and Products* (IBM), Document No. GC30–3072–0.

MATING THE LAYERS OF OSI AND SNA WILL BE BIG JOB

ERIC SACK

NEW YORK—One large area of concern has emerged recently for MIS departments faced with ever-expanding computer networks: the interconnectibility between Open Systems Interconnection (OSI) and Systems Network Architecture (SNA), International Business Machine Corp.'s reigning suite of data communications techniques.

Both architectures are "layered," that is, divided functionally into services that rely and build on those below. But differences in motive and impetus for creating the two have resulted in fundamental incongruities between the layered approaches.

"There's not a true mapability between the two systems," said Frank Dzubeck, principal of Communications Network Architectures Inc., a Washington, D.C.–based SNA consulting firm.

SNA—unwrapped in its general framework in 1974—was designed to enable IBM products to communicate in the most efficient way possible, insisting that the mainframe act as overseer of all events within the network.

OSI SPECS DEVELOPING
Meanwhile, the International Standards Organization (ISO) is slowly but relentlessly hammering out specifications for its Open Systems Interconnection, a reference model for data communications protocols among dissimilar devices from different manufacturers.

The International Standards Organization is comprised of representatives from participating countries, with decisions made on a consensus basis only. But Europe—where the need for the specifications is most accurate—is driving the effort.

"I think Europe will drive the United States," said Ronald Sander, president of the Sander Group Inc., a Severna Park, Md.–based SNA consulting house. "They're very hard-nosed about OSI over there, as they should be. They have tremendous protocol problems. You have to walk around with

a bunch of adapters just to use your electric shaver in different countries. That's not unlike the situation they have with computers," he said.

Pushed further by powerful interest groups in the United States—most notably General Motors and Boeing Computer Services with their OSI subsets—the OSI specifications are on the verge of becoming a reality in deliverable products, especially in key areas such as electronic messaging and document exchange.

THE DRIVE FOR STANDARDS
OSI sought from the outset to establish specifications that enable computer and communications equipment manufacturers to design protocols permitting the exchange of information in multivendor environments.

To achieve this, ISO chose a layered approach that is very different from that of SNA. The OSI model's design consists of seven layers, each of which contains a separate function. At least in spirit, these functions must be kept rigorously within that particular layer to ensure interoperability between vendors' equipment.

Implementation of the OSI software stack in one product must have a corresponding implementation in a device with which it attempts to communicate.

For example, if Transport Class 4—which provides error detection and recovery—is implemented in one device, the protocols must have peers at the same software layer in the other device for the function to operate.

Currently, many of the specifics of the fifth, or session, layer and the sixth, or presentation, layer of the OSI model are undefined. But functions at the seventh, or application, layer require these layers in order to operate.

Seventh-layer functions, such as the X.400 specification for electronic messaging, are currently being tested among several vendors. But these tests can only be achieved because all

Management Information Systems Week, October 6, 1986, 35. Reprinted with permission of MIS Week.

123

OSI	SNA
(L7) **Application**	**Application**

<table>
<tr><td>

- Identification of Partners
- Establish Authority to Communicate
- Privacy Mechanisms
- Cost Allocation/Quality of Service
- Error Recovery and Responsibility
- Select Dialog Discipline
- Information Transfer

</td><td>

Document Content Architecture

DOCUMENT	SNA
INTERCHANGE	DISTRIBUTION
ARCHITECTURE	SERVICES

</td></tr>
</table>

(L6) **Presentation**	**Function Mgmt. — NAU Services**

<table>
<tr><td>

- Data Transformation

- Data Formatting.

   —ASCII
   —Binary
   —EBCDIC
   —Numeric
   —Graphics

- Syntax Selection

</td><td>

- (SSCP), (PU), (LU)
- End User Access Authority
- Resolve Network Address/Net Name
- Select Session Parameters
- Management/Maintenance Services
- Sync Point Services
- Presentation Services Support
- Measurement Services

Function Mgmt. — Presentation

- Device Selection and Control
- Data Compression/Compaction
- Data Formatting

</td></tr>
</table>

(L5) **Session**	**Data Flow Chart**

<table>
<tr><td>

- Session Connection
- Session Establishment/Release/Sync
- Normal/Expedited Data Exchange
- Quarantine Service
- Exceptional Reporting

</td><td>

- Enforce Data Formats
- Enforce Chaining & Bracket Protocol
- Request/Response Correlation
- Assign Sequence Numbers
- Immediate/Delayed Response Modes
- Coordinate Send/Receive Modes
- Data Flow Suspension

</td></tr>
</table>

(L4) **Transport**	**Transmission Control**

<table>
<tr><td>

- **Establishment Phase:**
  Transit Set-Up, Network Service, Throughput,
  Delay
  Error Characteristics, Data Unit Size

- **Data Transfer Phase:**
  Blocking, Segmentation, Multiplex Connections,
  End-to-End Flow Control, Data Unit ID,
  Connection ID, Error Detection, Error Recovery,
  Expedite Data

- **Termination Phase:**
  Termination Reason, ID of Termination
  Connection

</td><td>

- Session Control:
  (Start), (Clear), (Resync)
- Connection Point Manager
- Message Unit Sequence # Check
- Encryption/Decryption
- Session Pacing
- Boundary Function:
  —Pacing
  —Data Traffic Protocols

</td></tr>
</table>

(L3)	Network	Path Control
• Network Address and End Point ID	• Path Selection
• Multiplex Net Connection to DLC	• Multiplex Net Connection to DLC
• Segmentation/Blocking	• Segmentation/Blocking
• Service Selection	• Transmission Group Control
• Select Service Quality	• Virtual Route Control with Pacing
• Error Detection and Recovery/Notify	• Explicit Route Control
• Expedited Data Transfer	• Route Extension Support/Peripheral
• Connect Reset w/Loss of Data/Notify	• Header Translation

(L2)	Data Link	Data Link Control
	• Data Link Connection	
	• Link Activation/Deactivation	
	• Map Data Units to Data Link Units	
	• Multiplex Data Units to Multiple Physical Connections	
	• Delimit Data Link Units	
	• Error Detection, Recovery and Notification	
	• ID and Parameters Exchange with Peer DLC Parties	

(L1)	Physical	Physical
• V.24 RS-232-C	• V.24 RS-232-C
• V.35	• V.35
• X.21, X.21 bis	• X.21, X.21 bis
	• S/370 Channel

Chart illustrates rough correspondence between the layered software approaches of Open Systems Interconnection and Systems Network Architecture. The chart suggests a one-to-one mapping of each of the layers, but in reality there is little correspondence between OSI network, transport and session layers on the left, and SNA's path control, transmission control and data flow control layers on the right. Apart from the mainframe channel attachment at SNA's physical level, there are many similarities between the bottom two layers of the models. Reprinted from "SNA Perspective," by permission of Communications Solutions Inc., San Jose, Calif.

vendors agreed to bypass the fifth and sixth layers in a kind of interim solution until those layers are better defined, according to Sander.

This interim solution illustrates the rigorous requirement for peer protocols in each layer of the OSI model.

In the case of electronic messaging, adding SNA to the picture seriously complicates networking and provides a glimpse at broader problems between OSI and SNA's layering techniques.

The seventh layers of SNA and OSI are roughly equivalent in function, in that both are designed to support actual user applications written on top of them. But similarities break down quickly because SNA's document distribution applications are provided by three sublayers of the top layer.

The topmost sublayer within SNA is a mainframe program that produces the document, which is in turn structured by Document Content Architecture (DCA), which is then passed under authority of the third sublayer, Document Interchange Architecture (DIA). The mainframe program requires DIA/DCA to generate the document to be distributed. IBM's implementation of DIA/DCA is dubbed DISOSS.

This sublayered document architecture operates effectively within SNA, and provides a rich assortment of directory services for documents. But for an SNA-based X.400 node, each protocol in SNA would require a like protocol in OSI with which to communicate—which won't happen because the protocols do not map one-for-one.

The alternative is to convert the DIA/DCA structure, going from the top down, into the X.400 structure, with not only loss of throughput but also some of the directory services of DISOSS that are not provided in X.400.

Also, DISOSS requires the functions of software below it, which must be converted to OSI protocols in order for the X.400 to make sense to the OSI node.

"HORRENDOUS GATEWAY"

"It's going to be a horrendous gateway between DISOSS and X.400," Sander said. "The higher you go in a software architecture, the more complex it is. The problem is that in order to have an X.400 document be exchanged with DISOSS, you're talking about messages starting at the top of one architecture and ending in the top of another.

"To do that, all the lower layers have to talk with counterparts. I suspect that's a monstrous technical challenge and will require something like a 4300 (IBM mainframe) to do the gateway," Sander said.

Difficulties with document distribution point to more fundamental differences between the layered approaches of SNA and OSI. In SNA, entities that control sessions and network paths reside in multiple layers and in multiple products, depending on the applications.

In the case of one particular mainframe application—Customer Information Control System (CICS)—LUs that support it are spread across the second, third, and fourth layers from the top. And the bottom part of the LU function is provided by Virtual Terminal Access Method (VTAM), Sander said.

LU 6.2 EXTENDS ACROSS LAYERS

Much of the SNA market discussion currently surrounds LU type 6.2, a software addressing platform that enables programs written on top of it to communicate with similar programs residing in other devices on a peer basis.

LU 6.2 is spread across several layers—a fact that was pivotal in the European Computer Manufacturers Association's (ECMA) decision to reject LU 6.2 as an OSI application-layer protocol, many industry watchers have said.

"It would have been a complete refutation of the layered approach of OSI, where each layer stands by itself," Sander said.

These few examples show that IBM, when it decides to interconnect to the OSI world, will do it with gateways rather than the simpler bridges. A bridge occurs when two roughly equal entities can communicate after getting over a physical incompatibility such as transmission medium, or a logical restraint such as network addressing boundaries.

Gateways, on the other hand, require conversion of higher-level structures, and therefore impact throughput.

"In the long run, you'll have to use gateways to solve needs between SNA and OSI," Dzubeck said. "If you have the need, then there's no restriction except for throughput."

Sander said that gateway throughput would be unacceptable for certain applications. "In an application that needs less than two- to four-second response time, that gateway is going to be a bottleneck," he said.

8 LOCAL AREA NETWORKS

MINICOMPUTERS
LOCAL AREA NETWORKS
    Bandwidth
    Topology
    Access Protocols

PRIVATE BRANCH EXCHANGES
CLARK FIBERS, INC.

AFTER READING THIS CHAPTER, YOU WILL UNDERSTAND THAT:

- Because of networking, we are on the edge of a communications revolution.
- Minicomputers can tie together isolated PCs so that information is shared.
- Local area networks can also connect isolated equipment.
- In some situations, office telephone system can provide local data communications.

When process engineers, data-processing professionals, customer service staff, quality assurance people, and others are linked through networking, the result is improved quality, responsiveness, and productivity. Instead of isolated islands of technology, we have integrated information on a plant-wide basis: the factory floor and the front office are unified, and process control computers communicate with corporate accounting computers. In other words, a successful business operation has to function as a single entity in order to provide the customer with quality products at the lowest cost. Networking promotes this.

Accordingly, we are on the edge of a communications revolution. Putting the right information in the hands of the right person at the right time to make the right decision is becoming a requirement, and timely access to information is the key.

In many areas of business the 80/20 rule is useful. For example, 80% of the sales volume is made to 20% of the customers, or 80% of the value of inventory is represented by 20% of the items.

The 80/20 guideline applies to networking as well: 80 to 90% of the information generated locally is used locally and only 10 or 20% is needed

FIGURE 8–1 ————————————————————————————————————

Information distributed within the local establishment.

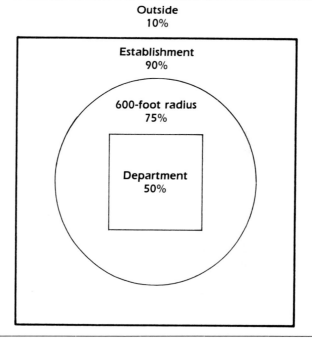

An Introduction to Local Area Networks (IBM, November 1983), Publication No. SC20–8203–0, 2–3. Courtesy of International Business Machines Corporation.

elsewhere (see Figure 8–1).[1] This means that a manufacturing plant, hospital, or college has information needs that are predominantly local — a good reason to emphasize local information handling.

Most information exchanges occur over relatively small distances: e.g., within a plant or a campus. Most work in a local environment is organized into groups of people doing related jobs, and this creates the need to tie computers together at the group level (i.e., within each department). The primary benefits from having the equipment communicate are:

- information moves faster, with more flexibility, and allows improved responses to the unexpected
- easier access to information, which results in better decision making
- sharing information and related resources (e.g., printers, disk drives).

Since there are an estimated four million departments and other work groups in U.S. companies and government agencies, many people are seeking local means for improving access to information.[2] Three solutions have emerged, all of which are complex mixtures of computing, communications, and software:

1. minicomputers to tie together isolated personal computers
2. local area networks to connect the equipment
3. in some situations, the telephone system to provide local data communications.

MINICOMPUTERS

Minicomputers (or minis) can tie together isolated personal computers (PCs, or microcomputers) so that information is shared. Also, they can connect the PCs to the organization's mainframe. Minicomputers are smaller than mainframes and bigger than PCs; they can thus help handle departmental processing by allowing the PCs both to share information and to communicate with the central computer.

This application of minicomputers may disappear as PCs become more powerful, with the software and the communications capacity to talk directly to mainframes. Local area networks (LANs) would then be the best solution. Critics believe, however, that LANs still do not work well enough, noting that LANs still lack the software to allow groups of PCs to work on the same problem (i.e., with the same data) simultaneously. Further, minicomputer proponents argue that LANs do not yet tie computers into a cohesive information system.

—— LOCAL AREA NETWORKS ——————————————

Our telephone system developed to serve our needs for voice communication. As data communications emerged, the digital data was modified by modems to travel over this analog voice system. The telephone network is thus used to transmit data over relatively long distances.

As low-cost, 16-bit personal computers became common, however, businesses discovered that the bulk of their communications were over relatively short distances. Personal computers provided an economical means to automate such short-range internal communications: the local area network. In contrast to the long-distance telephone network, LANs are used only for local data communications.

Examples of LAN applications include:

- office personal computers connected to shared word-processing and data base files
- point-of-sale and inventory transaction systems (see "Data Bases" in Chapter 5)
- robots and machine controllers communicating with computers on the plant floor (see "GM Maps the Future" at the end of this chapter).

The term *local area network* is used both for a communication technology and a multiuser computer installation. As a communication system, LANs should conform to the ISO seven-layer model (see Chapter 7). The objective is to allow users with different types of equipment to plug directly into the network without having to worry about compatibility. LAN architecture includes the bottom two ISO layers (see Figure 7–2):

- the physical layer, which includes the cable and connections to the cable so devices can send and receive (see Figure 8–2)
- the data link layer, which defines the protocol that devices use to gain access to the cable for transmission, as well as the data packet size and format (this includes such functions as addressing, error recovery, and flow control).

National and international organizations are working on LAN standards to insure compatibility among devices and trouble-free data communication. The standards are for commercial and light industrial applications in areas such as word processing, digital voice transmission, file transfer, data base access, and electronic mail.

While local area networks are part of the overall communication network, and often provide a gateway to common carriers, they have several characteristics that distinguish them from long-distance voice networks:

FIGURE 8-2

Transmission media.

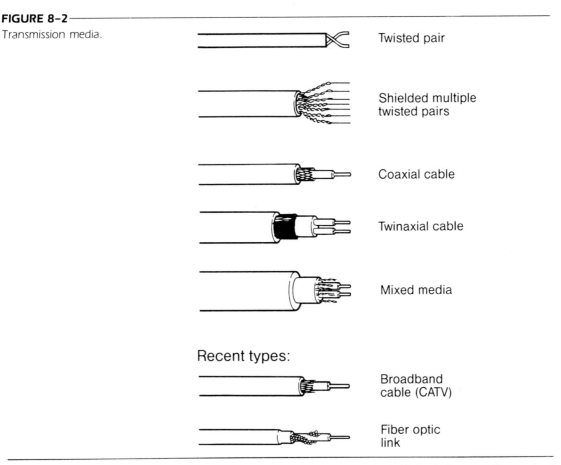

Twisted pair

Shielded multiple
twisted pairs

Coaxial cable

Twinaxial cable

Mixed media

Recent types:

Broadband
cable (CATV)

Fiber optic
link

Local Area Networks (IBM, 1983), Publication No. G320–0108–0, 10. Courtesy of International Business Machines Corporation.

- LANs service a limited geographic area, such as a manufacturing plant or corporate headquarters
- LANs use a privately owned and controlled communication medium
- LANs operate at relatively high speeds (greater than 125,000 characters per second), in contrast to typical long-distance networks (120–240 characters per second).

As multiuser computer installations, LANs are an alternative to the traditional CPU with its connected terminals and its multiuser operating system. While both systems can share resources such as printers and disk drives, the terminals connected to a CPU are sometimes "dumb" and cannot act as independent, stand-alone computers or execute their own application programs. Each LAN station has to be able to operate independently, and so must be intelligent.

Figure 8–3 illustrates one possible LAN. The person operating PC 1 prepares a report (i.e., a word-processing or text file), which is then stored on a file server. The individual using PC 2 can access and edit the report, and the operator of PC 3 can then print the edited report on the shared printer (attached to PC 1) and save the file on the shared hard disk (attached to PC 1).

As noted, communications over the telephone network occur at relatively low speeds and tie up a communication channel continuously for the duration of the call. In contrast, LAN file transfers (e.g., moving a file from the file server to a personal computer in the network) occur

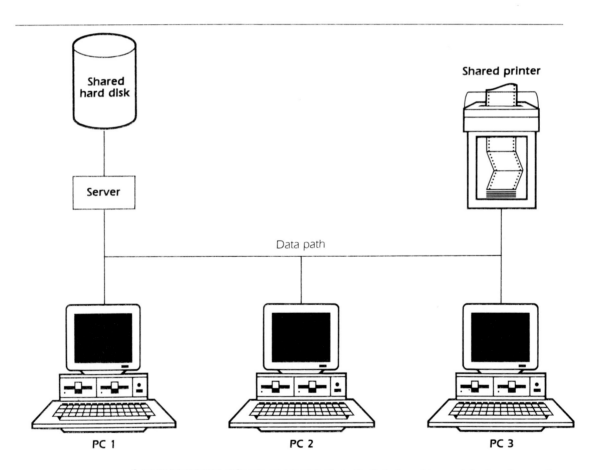

Shared hard disk

Shared printer

Server

Data path

PC 1

PC 2

PC 3

A server is hardware, software, or a combination of both that provides specialized services to other network devices. It can be part of a workstation or it can be a separate device. For example, a disk server makes access to a disk drive as quick and efficient as possible.

FIGURE 8–3   Local area network.

relatively fast and do not tie up the channel. In fact, LAN data transfers occur about as fast as data movements inside some central processing units.

LANs may have to handle demands to transmit, access, store, and print data from several stations nearly simultaneously. To provide a fast, orderly, error-free information flow, data streams are formatted into packets (similar to the SDLC format in Figure 6–4) for transmission according to network protocols. Each personal computer (or minicomputer) has an adapter card that provides the LAN interface, and the adapter card contains the logic chips needed to send and to receive the packets.

As usual, software is crucial. For example, personal computers connected to shared disks and printers need special operating systems to deal with such matters as assigning priorities to users and handling multiple requests to access the same file. Each local area network supports a specific operating system, which may place restrictions on the PC; for instance, the operating system may not permit it to be used locally while simultaneously responding to a remote request for access to its attached printer or file server. Further, preferred application programs (e.g., electronic mail and data base packages) may not run on that operating system.

Three other important factors that bear on a LAN's cost and capabilities are data path capacity (bandwidth), geometric structure (topology), and access control (protocol). These are discussed next.

Bandwidth

Bandwidth describes the LAN's ability to move data, and the way this capacity is used. There are two different transmission technologies that provide this function: broadband or baseband.

Broadband transmissions

Broadband systems use a relatively broad range of frequencies, providing multiple communication paths; in other words, a broadband LAN transmits over a broad range of frequencies.

As noted in Chapter 2, analog signals are continuous waves of either voltage or current amplitude. Information may be transmitted by varying either the amplitude or the frequency of the continuous waves (see Figure 8–4). Telephones use analog signals: the electrical signal varies proportionally with the sound waves entering the mouthpiece of the telephone, and at the other end of the line, the continuously varying electrical signal is converted back into sound waves. Another example of analog technology is a cable television network, where several frequencies are transmitted over the same cable. The user can tune to any of the frequencies by selecting a channel.

FIGURE 8-4 ————————————————————————————————————

Analog signals.

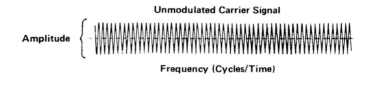

Amplitude

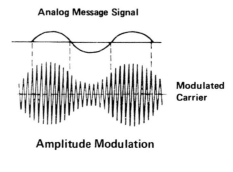

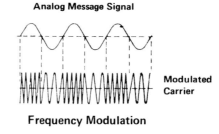

An Introduction to Local Area Networks (IBM, 1984), Publication No. GC20-8203-1, 3-6. Courtesy of International Business Machines Corporation.

In much the same way, broadband systems allow many devices to share a cable by assigning specific transmission frequencies to each device connected to the line. This technique is called **frequency division multiplexing** and allows a number of separate transmissions to take place simultaneously on a single cable (see Figure 8–5). If two LANs used the same broadband cable, for example, the devices connected to one LAN would use one set of frequencies and those connected to the other LAN would use a different set of frequencies.

Each device on the network requires a *radio frequency modem* (or RF modem) to send or receive on its assigned frequency (see Figure 8–6). Devices on different frequencies cannot communicate without additional frequency translation equipment, such as switchable or multifrequency RF modems. The *head end* is a component of a broadband network that allows devices to send and receive on a single cable. Channels may be shared,

FIGURE 8–5 ————————————————————————————

Broadband
systems—frequency
division multiplexing.

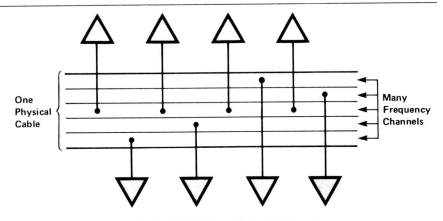

Stations With Unique Dedicated Channels

An Introduction to Local Area Networks (IBM, 1984), Publication No. GC20-8203-1, 3-8.
Courtesy of International Business Machines Corporation.

dedicated (permanently allocated to a specific purpose, such as high-volume data transfer), or switched (temporary connections that allow two devices to communicate).

Video, voice, text, and data can all be supported by broadband systems. Dedicated and switched channels remain tied up for the duration of the transmission, so can be used for voice or video as well as data. Shared channels require some method for access and control (see the "Access Protocol" section below) and are not normally used for voice or video.

Major cost factors in broadband systems are the variable-frequency, high-speed modems and the relatively complex interface units (the adapter

FIGURE 8–6 ————————————————————————————

Broadband
systems—modems.

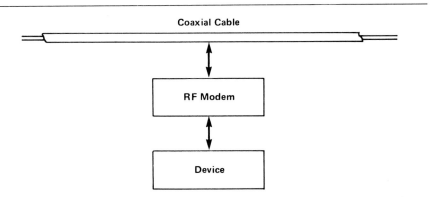

An Introduction to Local Area Networks (IBM, 1984), Publication No. GC20-8203-1, 3-8.
Courtesy of International Business Machines Corporation.

FIGURE 8-7

Transmission technique and media.

TRANSMISSION TECHNIQUE	MEDIUM
baseband (digital)	copper wire • shielded twisted-wire pair • coaxial cable • CATV cable fiber optics cable
broadband (analog)	copper wire • coaxial cable • CATV cable

An Introduction to Local Area Networks (IBM, 1984), Publication No. GC20-8203-1, 3-37. Courtesy of International Business Machines Corporation.

cards which connect each PC to the network). These costs may make broadband technology unsuitable for low-cost personal computer LANs.

Figure 8-7 compares transmission media for broadband and baseband transmissions.

Baseband transmissions

Baseband systems use digital technology, as do computers themselves. Baseband technology allows several devices to share a cable by means of a time-sharing technique called *time division multiplexing:* each device is assigned a specific time slot—a few thousandths or millionths of a second—in which to transmit. Only this device can transmit in this time slot (see Figure 8-8).

In a baseband system,

- many different types of transmission media can be used; for instance, twisted-wire pairs, coaxial cable, and fiber optic cable (see Figure 8-2)
- there must be a protocol to assign time slots and control access to the network (see below)
- modems are not required to connect devices to the network
- any number of devices can be used to transmit or receive information (without complex additional equipment), as long as there are enough slots available
- data transmission rates of 3-10 Mbps over distances of one to three miles are typical
- only data and text can be transmitted (voice and video are seldom supported)
- transmission is all-digital and half-duplex (a node can either send or receive, but can't do both simultaneously)

FIGURE 8-8

Time division
multiplexing (TDM).

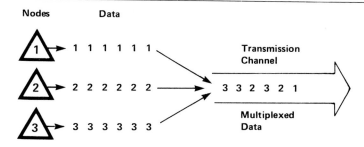

An Introduction to Local Area Networks (IBM, 1984), Publication No. GC20-8203-1, 3-10.
Courtesy of International Business Machines Corporation.

- there is no central power supply; the power comes from the attached equipment.

In general, baseband technology tends to be less expensive than broadband systems. Installation, device interfaces, and traffic control are usually cheaper for digital networks.

Topology

A local area network has to be physically connected in order to distribute data locally among multiple users, and the geometric view of how the network is connected together is called its *topology*. In other words, topology means the physical layout of the transmission medium—the wires or cables—that connects devices in the network. The points where devices connect to the medium are called **nodes.** Topology affects (among other things) how reliable and readily expandable a LAN may be.

Most LANs use a *broadcast* topology, wherein every message goes to every node, and an individual node only recognizes and acts on messages flagged with its address. The most common broadcast topologies are *bus* and *ring* (see Figure 8-9), and a frequently used nonbroadcast topology is *star*:

- ■ *Bus.* A bus is a length of cable not connected in a loop or ring. Individual devices tap into it and a device or node broadcasts its signal in both directions to all other nodes. No routing decisions are required. A bus system normally gets its power to transmit data from the hardware attached to it, not from a separate power system. Some examples of bus topology are Ethernet and Wangnet.
- ■ *Ring.* Messages from a transmitting node flow in one direction around the ring, passing from node to node and being regenerated at each node. No routing decisions have to be made. Ring topology is thus circular and unidirectional.

FIGURE 8-9

Typical LAN
topologies.

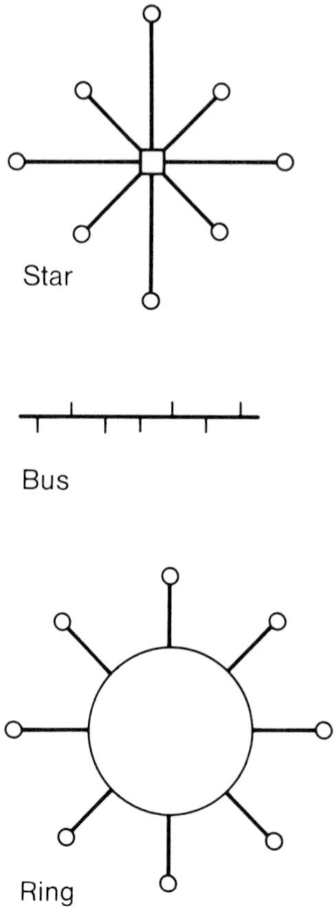

Star

Bus

Ring

Local Area Networks: A Review (IBM, 1983), Publication No. G320-0108-0, 12. Courtesy of International Business Machines Corporation.

■ **Star.** All transmissions pass from the transmitting node through a central controller to the receiving node. This central controller thus manages all communications. Telephone systems use star topology, with the handsets at the nodes and the private branch exchange as the controller.

Figure 8-10 shows some advantages and disadvantages of bus, ring, and star topologies.

Access Protocols

A local area network *access protocol* decides who gets access, and when. Protocols control access to the transmission medium from a connected

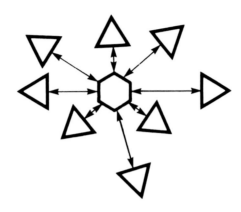

STAR TOPOLOGY

ADVANTAGES	DISADVANTAGES
• cable is easily modified to accommodate changes and moves • defective nodes are easily detected and isolated	• uses the most cable • failure of the central controller disables the entire network

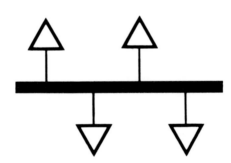

BUS TOPOLOGY

ADVANTAGES	DISADVANTAGES
• uses the least amount of cable because it provides the most direct cabling routes	• requires a means of bidirectional traffic control along the bus to prevent simultaneous transmissions • a break in the cable cannot be easily detected

FIGURE 8–10   Star, bus, and ring topologies compared.

An Introduction to Local Area Networks (IBM, 1984), Publication No. GC20-8203-1, 3–26. Courtesy of International Business Machines Corporation.

FIGURE 8–10
(Continued)

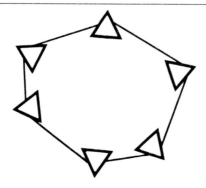

RING TOPOLOGY

ADVANTAGES	DISADVANTAGES
• a break in the cabling can be readily detected	• modifying the ring to accommodate changes may be difficult
• communication control is simplified because the ring is unidirectional and a closed circuit	• uses more cable than bus topology, but less than star
• more distance can be covered by the network because each node regenerates the signal to full strength	
• lends itself to the use of fiber optics because of its unidirectional nature	

node, and how access is granted depends on whether the LAN control method is *centralized* or *distributed*, as illustrated in Figure 8–11. In addition, access to the network may either be by *contention* or *guaranteed*, which results in four different access protocols:

Circuit switching protocol uses centralized control (characterized by the star topology) and contention access. Telephone systems use circuit switching to establish communication paths: when a node (e.g., telephone) demands (or contends for) access, the central controller (e.g., a private branch exchange) connects the calling node to the called node If the called node is already busy, access is denied. Otherwise, a dedicated connection is made, which is unavailable for any other use until the two nodes complete their communication.

Polling protocol also uses the star topology, but access is guaranteed rather than by contention: the central controller polls each node in a predetermined order, in effect asking the node whether it seeks access. If not,

FIGURE 8–11 ————
Access Methods/
protocols.

CONTROL METHOD	ACCESS METHOD	
	CONTENTION	GUARANTEED
centralized	circuit switching	polling
distributed	CSMA/CD	token passing

An Introduction to Local Area Networks (IBM, 1984), Publication No. GC20–8203–1, 3–28.
Courtesy of International Business Machines Corporation.

the controller polls the next node. Whenever a node seeks access, the controller determines the best transmission path and routes the message before it returns to its polling schedule.

CSMA/CD (carrier sense multiple access/collision detection) is the access method frequently used in bus topologies: it has distributed control and contention access. Before initiating a transmission, a node listens for a carrier signal indicating that the bus is in use. If it is in use, the node waits and tries again. If the bus is idle, the node sends its message and listens for a collision with another message. If it doesn't hear anything, the node assumes that the message has arrived. If there is a collision, the node can detect it and transmit an error signal to all the other nodes before starting over again (see Figure 8–12). Contention access protocols like this one are fine for LANs with many nodes and limited communication needs: simply transmitting as needed is more efficient than implementing a complex control method. But a LAN with many active nodes would tend to have high message collision rates, and since each collision takes time to resolve, this might result in a drop in transmission volume during active periods.

Token passing protocol (distributed control, guaranteed access) is often used with both ring and bus topologies. A *control token* (a number of bits that control transmission in the LAN) is passed from node to node. Only the node with the token may transmit; all others listen. If the node has nothing to transmit, it passes the token on to the next node; if it does have a message, it attaches it to the token for transmission (see Figure 8–13). In a ring topology, the token is always passed to the next node in the ring. In a bus topology, each node is programmed to pass the token on to a specific node, which results in a logical ring (see Figure 8–14).

———PRIVATE BRANCH EXCHANGES———

The private branch exchange was developed for voice communications, but in some situations the PBX (or its descendants, the private automatic branch exchange and the digital PBX) can effectively provide LAN

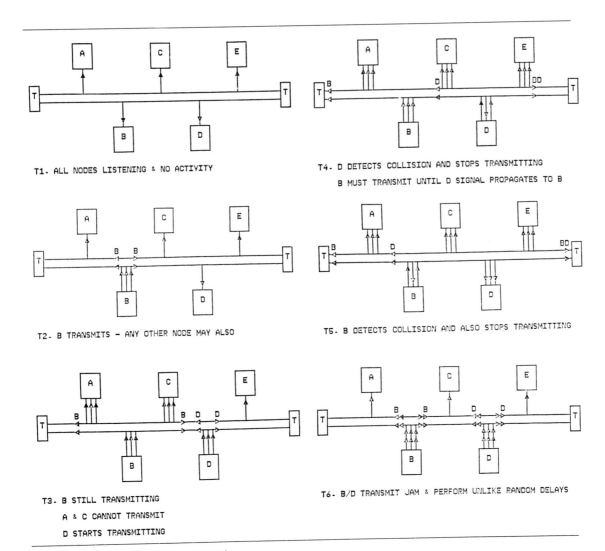

FIGURE 8-12 CSMA/CD bus operation.

Local Area Network Concepts (IBM, August 1984), Publication No. G320-0161-0, 34-36.
Courtesy of International Business Machines Corporation.

capabilities. One benefit is that the PBX uses existing telephone wiring, so the expense of stringing additional cables can be avoided. A PBX uses circuit switching protocol: star topology, centralized control, and demand access.

Whether a PBX will be practical for a LAN depends on the mix of voice and data traffic (where the traffic is mostly voice, a separate data network

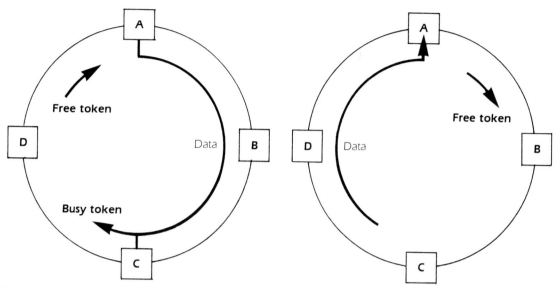

In this example, node A is ready to send data and looks for a free token. Node A changes the free token to a busy token and appends the data to the token, addressing it to node C. Node C receives the busy token and copies the data addressed to it. Node A then removes the busy token and data from the ring, and generates a free token to replace the busy token.

FIGURE 8–13 Token passing ring.

Local Area Network Concepts (IBM, 1984), Publication No. G320–0161–0, 39. Courtesy of International Business Machines Corporation.

may be advisable), and on the type of data traffic: batch, interactive, or message:

■ **Batch** traffic usually involves the transmission of large volumes of data between high-speed devices, so dedicated point-to-point channels (such as leased lines) may be best.

■ **Interactive** traffic normally requires fast response times, and a PBX may not have the network switching capacity needed to handle this.

■ **Message** traffic, such as electronic mail, is well-suited to PBXs because the messages tend to be short and the transmission speed is slow.

The digital technology used in VLSI chips has been adapted to PBXs, which results in a digital PBX (see Figure 8–15). Its all-switching system allows the integration of voice and data traffic, and uses a codec (coder-decoder) to convert voice into digital signals for switching. In certain situations, digital PBX may be used instead of a LAN.

Much of our working lives is spent communicating with one another, but, until recently, these communications were rarely automated. In 1984,

FIGURE 8-14
Token passing.

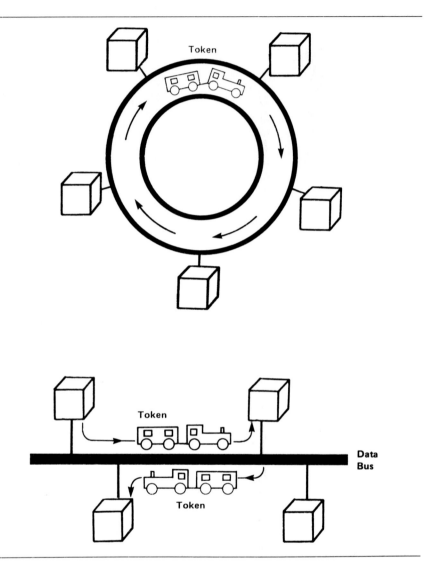

An Introduction to Local Area Networks (IBM, 1984), Publication No. GC20-8203-1, 3-32, 3-33. Courtesy of International Business Machines Corporation.

there were an estimated four million personal computers in the United States, one million of which (25%) were networked. In 1989, it is estimated that there will be 25 million personal computers, 19 million of which (75%) will be networked.[3] Local area networks will be one of the ways in which organizations will be able to process timely information and distribute it among the many devices on their premises.

FIGURE 8–15 ———
PBXs.

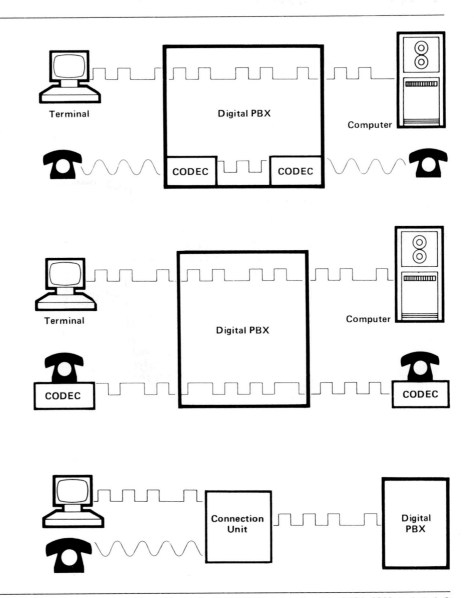

An Introduction to Local Area Networks (IBM, 1984), Publication No. GC20–8203–1, 4–6, 4–8. Courtesy of International Business Machines Corporation.

CLARK FIBERS, INC.

The case study of Clark Fibers, Inc. at the end of this chapter will help the reader learn about LANs and their implementation. As you read the case, keep the following questions in mind:

1. What are the main problems?

2. What are possible solutions?

3. Which solutions would you recommend? Why?

SUMMARY

1. Because of networking, we are on the edge of a communications revolution.

2. Eighty to 90% of the information generated locally is used locally— in the same plant, hospital, campus, etc.

3. Three solutions for tying local equipment together are minicomputers, LANs, and PBXs.

4. Minicomputer proponents believe that LANs do not yet tie computers into a cohesive information system.

5. The term LAN is used both for a multiuser computer installation and a communications technology.

6. Bandwidth describes the LAN's ability to move data, and the way this capacity is used.

7. The two transmission technologies used by LANs are broadband (analog) and baseband (digital).

8. Topology is the physical layout of the medium that connects the devices in a network. The three most common are ring, star, and bus.

9. LAN protocols decide who gets access, and when. Control may be centralized or distributed, and access may be guaranteed or by contention.

10. Circuit switching and polling are common centralized protocols.

11. CSMA/CD and token passing are common distributed protocols.

12. In certain situations, a PBX or digital PBX may be used instead of a LAN.

REVIEW QUESTIONS

1. Why are we on the edge of a communications revolution?

2. What are the primary benefits of having equipment communicate?

3. Discuss the three solutions that have emerged for automated local communications.

4. Describe how minicomputers are used to tie together departmental PCs.

5. Give two or three examples of LANs in an office or plant.

6. How do LANs conform to the ISO model?

7. Name some characteristics that distinguish LANs from long-distance voice networks.

8. Why do LANs use packets?

9. What are bandwidth, broadband, and baseband?

10. What are ring, bus, and star topologies? What is the importance of each?

11. What are LAN protocols?

12. Discuss CSMA/CD and token pasing.

13. If you're considering a PBX for local data communications, what factors should you examine?

14. Define PABX, codec, and digital PBX.

15. Why does GM use MAP? (See the article beginning on page 148.)

ENDNOTES

1. These estimates are from *An Introduction to Local Area Networks* (IBM, November 1983), Publication No. SC20–8203–0, 2–3, and Juan Buines, "Local Area Networks," *Telecommunications for Management,* ed. Charles T. Meadow and Albert S. Tedisco (New York: McGraw-Hill, 1985), 143.

2. See "The New Computer Wars," *Business Week,* July 15, 1985, 97.

3. "The World on the Line," *The Economist,* November 29, 1985, Telecommunications Survey, 7.

CLARK FIBERS, INC.

In late February 1983, Peter Lowell, vice president of Clark Fibers' management information department, was considering a proposal to spend $500,000 immediately to install a local area network (LAN) for intracompany communications.

Clark Fibers' R&D employees would be the first to use the network extensively. Both President Michael Lyons and Chairman Edward Reardon were committed to a strong R&D department and increased investment in that area. The LAN technology was new, however, and the costs were high. Lowell wondered whether he should wait until the technology was more mature before making such a large investment in it. While the projected cost savings were impressive, they were soft savings such as improved productivity, for which the company would have to spend hard dollars. If he did approve the purchase, Lowell wondered, what would the system's ultimate impact be?

THE COMPANY

Clark Fibers, Inc. had 1982 sales of $900 million, with net earnings of $96 million. The company employed approximately 10,000 people in more than 120 plants in the United States and eight other countries. As shown in Exhibit 1, the firm was organized into three operating groups: industrial services, engineering, and fibers, and one relatively autonomous subsidiary; Resins, Inc.

All of the line operating groups reported to the president. With the exception of Lowell's management information department, all staff functions reported to the vice chairman. A central management committee, which met weekly, oversaw the budgeting process and approved all major pricing and capital expenditure decisions. Once the budget was approved, the product group vice presidents had considerable latitude to make operating decisions.

Corporate organization*

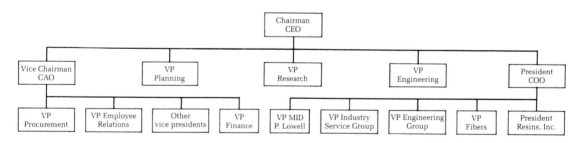

*Depicts reporting relationships only.

EXHIBIT 1 *Corporate organization*

This case was prepared by Leslie R. Porter and John Sviokla as the basis for class discussion rather than to illustrate either effective or ineffective handling of an administrative situation. Reprinted by permission of the Harvard Business School.

Clark Fibers' central corporate facility, where the LAN would be installed, was located at Pittsfield, Massachusetts, approximately 70 miles from Boston. It housed the headquarters of the line divisions, the main R&D facilities, and the corporate offices. This campus, as it was called, included eight major buildings, one for manufacturing, five for administration, and two for R&D. (See Exhibit 2.)

The R&D department was responsible for developing products and processes for all divisions, and served as a resource for the entire com-

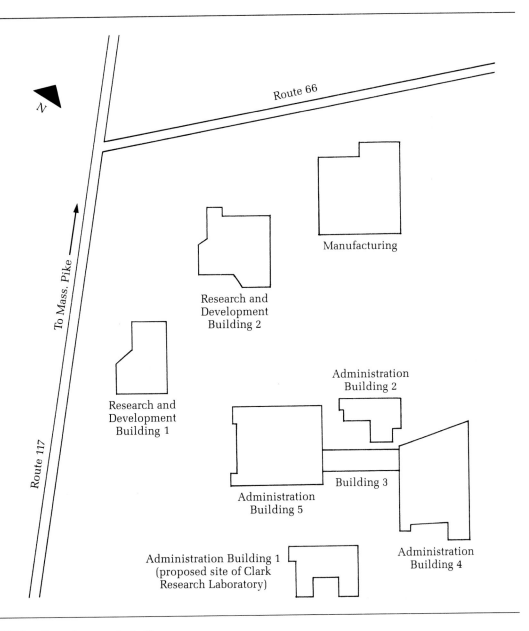

EXHIBIT 2 *Central corporate facility.*

pany. Because Clark Fibers expected much of its growth to come from new products, top management viewed R&D as critical to the company's future success.

THE MANAGEMENT INFORMATION DEPARTMENT

Senior management at Clark Fibers had long recognized the importance of an efficient, well-coordinated data processing (DP) effort. For this reason, DP planning, development, operations, and support activities were concentrated in the corporate management information department (MID). Peter Lowell reported directly to the president. In 1982, the department's budget was approximately $20 million, two-thirds of which was spent on domestic activities; the remaining costs were incurred primarily by parallel organizations in Europe.

Within MID two managers reported to Lowell directly. (Exhibit 3 shows the MID organization chart.) The development and support groups— business systems, office automation (OA), scientific information systems (SIS) and management systems—reported directly to Ron Fleming, assistant director of MID. The facilities groups—data entry, technical support, and computer operations—reported to Brian Park, director of facilities.

Facilities, SIS, and OA are all involved in the local area network decision.

Facilities. The facilities group was responsible for operating and maintaining all of Clark Fibers' computing hardware and operating systems and monitoring the use of equipment. Its primary concerns were the system's overall performance, reliability, and integrity.

In 1981–1982, facilities was engaged in several major projects including a major software conversion, addition of new remote locations, and expansion of the on-line mainframe terminal network.

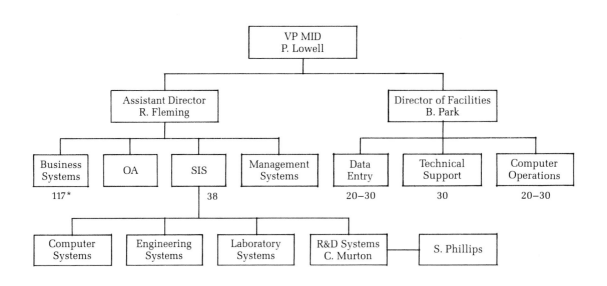

*Number of people

EXHIBIT 3 Organization chart.

Brian Park commented on facilities' relationship with the user community:

> Many of the newer applications come from people outside the facilities group. Time sharing, interactive graphics (CAD/CAM), and personal computers were all initiated by other groups within MID. The key issue we continually deal with is, how we can bring change into an organization that is largely concerned with the day-to-day operation of a large MID plant. Often a user group acts as a change agent: they start up a project, then after it gets going, it folds into the facilities area so that it can be professionally managed and we don't end up with many separate systems.

Scientific Information Systems (SIS). SIS's mission was to provide information systems for Clark Fibers' engineering, scientific, and R&D communities. These departments included many sophisticated users of information technology who were able to set up, maintain, and operate their own equipment. Ordinarily, however, SIS developed, implemented and maintained new software applications; facilities was responsible for the computer hardware and operating systems on which the programs ran; and the users ran the programs.

SIS comprised four groups: process systems, engineering systems, laboratory systems, and R&D systems. Over time, the R&D group had implemented a number of systems to improve productivity, such as process control, computer-aided design, graphics and simulation. These efforts had not been well coordinated, however, and the result was a large variety of minicomputers and peripheral equipment throughout the company.

Craig Murton, manager of R&D systems, had joined Clark Fibers' R&D department 15 years earlier, with a strong interest in computer automation. In early 1980, he decided he could implement automation more effectively if he moved into the MID organization. With the approval of all, he joined SIS as manager of the then new R&D systems group. Shortly thereafter, he hired Steve Phillips as assistant manager.

Office Automation (OA). The office automation group, under the direction of Mark Hatfield, was responsible for investigating new office products and alternative ways to improve office productivity at Clark Fibers. The group was primarily concerned with assessing new products and bringing them into the company. Mark Hatfield explained:

> OA is an agent of change. Our job is to scan the marketplace and bring the best products in house, get them evaluated, and implemented, and take them through the initial stage. Then we bow out and pass responsibility for operation to the user group and the facilities people.

To do this effectively, OA had to integrate the needs of the end users and the people who would eventually maintain and operate the systems. In addition, many of the solutions OA suggested, such as word processing and electronic mail, crossed many functional boundaries. Consequently, OA personnel saw themselves as facilitators who balanced the wishes of many constituencies while keeping current on new vendor offerings and products.

COMMUNICATIONS

The communications department, a staff group outside MID, was responsible for advising the line groups on various types of communication, such as voice, video, audiovisual, data, and reprographics. Hal Oppenheimer, manager of the telecommunications group within the department described the relationship of the communications department to the office automation group:

> Communications uses OA as internal consultants. They are sort of an R&D arm to look at new technologies. Here at Clark Fibers, we like to run a lean staff organization, so we often need to draw on resources outside our department, like OA, to get the job done.

HISTORY OF THE LAN ISSUE

Since 1979, several organizations in Clark Fibers had been interested in local area networking as a possible solution to their communications problems. In the winter of that year, the communications department and MID jointly commissioned a consultant to study the issue of interconnecting the data, voice, video, and computing technology on campus.

The interconnection issue was complicated by the widespread use of office equipment built by different vendors. The varied base of equipment and an active vendor market made flexibility a key criterion for any solution. At the same time OA wanted to make the office of the future into the office of the present as soon as possible. To them, the LAN option seemed a promising means to achieve resource sharing and better communications which would also lay the groundwork for greater office automation.

The communications department was interested in the LAN issue for several reasons. Currently, computing and office equipment was connected to remote areas over phone lines, using modems; this method required heavy telephone use. As the number of lines on campus increased, communications began to think about buying a Computerized Branch Exchange (CBX) to increase phone efficiency and capacity. They felt responsible for the interconnection of a CBX to any other communications media at the company. Also, some of their staff had experience in some facets of the LAN technology.

At this stage, Craig Murton entered his new role as manager of R&D systems. He commented.

> After I moved into my new job, I began to concentrate on how to improve productivity in the R&D area. It became clear that better information technology would be essential and that a versatile communications capability would be a key component.

Just as a wide variety of equipment had been installed in Clark Fibers' offices, the company's laboratories employed myriad minicomputers, microcomputers, and process control equipment, all from different vendors. This diversity further complicated the time-consuming and sometimes tedious job of setting up experiments and gathering information. Steve Phillips described the situation:

> If an engineer or chemist wants to get access to graphics, minicomputers, and the mainframe—a typical combination of needs for an experiment—he has to put three terminals in his office, one for each system. It is not uncommon to see a $40,000-a-year professional lugging a terminal down the hall to hook it up to an experiment to get the data he needs.

Murton and Phillips began searching for solutions. At first they considered installing a huge switchboard, much like a 1920's telephone exchange, with an operator to patch the necessary equipment together by hand. Phillips' financial analysis showed this option was not feasible, however. Furthermore, it did not solve the problem of equipment incompatibility and would only add to the "spaghetti factory" of cables already in place. Both Murton and Phillips felt the LAN technology would be worth a test.

THE CLARK RESEARCH LAB AND THE WORKING COMMITTEE

In 1981 Clark Fibers was planning to renovate Administration Building #1 for use as an R&D facility. The proposed facility was to be named the Clark Research Lab, in memory of the company's founder. Murton and Phillips now had to determine what type of LAN, from what vendor, should be installed. Phillips undertook a preliminary literature and vendor survey to familiarize himself with the product offerings and the important technological issues in choosing and running a LAN.

By mid-1981, Murton had learned of the consultant's study and Mark Hatfield's efforts in OA as well as the growing interest in the new technology around the campus. Through discussions, he and the other managers determined who would be directly affected by the LAN. Murton then set up a working committee that included Hal Oppenheimer from the communications department, Eileen Sawyer from office automation, Steve Phillips from SIS, Archie Wolf from facilities, and two representatives from purchasing.

The committee's long-term mission was to develop a coherent corporate approach to local communications. In the short term, it was to recommend a particular type of LAN for installation in the Clark Research Lab. The most pressing need was to determine the core technology of the LAN, so that the appropriate cable could be installed in the laboratory as it was being built. February 1982 was the deadline.

The first few committee meetings were tumultuous. As Eileen Sawyer commented:

Actually, our first goal was just to get everyone to sit down in the same room and discuss the issues. That was quite a feat in and of itself, and even so, several memos documenting various opinions were getting a broad distribution. There was a lot of frustration over four basic issues: role definition, resource availability, priorities, and perception of alternatives.

The committee members had diverse backgrounds and differing levels of expertise in LAN technology. In an effort to build consensus, the committee first educated itself about the critical issues involved in looking at a LAN. Phillips attended a number of educational sessions and vendor information presentations to learn about the basic technologies, and then shared this information with the others. Mark Hatfield asked the consultant to brief them on the state of the art and to describe his findings about Clark Fibers' needs.

LOCAL AREA NETWORKS OPTIONS

In 1981, over 25 vendors were offering LAN products. Local area networks are designed to be run and maintained by their users. These systems typically offered faster data transfer than a longhaul communications network, but over a limited geographic area. The Clark Fibers' working committee looked at many possible options and quickly narrowed their focus to three basic types of technology: CBX, baseband, and broadband.

A CBX switches data communications from point to point, just as it does phone communications. It would operate primarily over Clark Fibers' existing twisted pair phone wires, therefore requiring a minimum of cable installation. The committee found that the best digital CBXs available would satisfy most, but not all, of the company's data needs. There were promises of better data rates and more capacity to come, but only later.

A baseband system, which provides a single digital channel with relatively high data capacity, would have answered many of Clark Fibers' current data needs in both R&D and the office. In comparison with broadband technology, which provides many channels, with analog or digital signals, a baseband system is simpler in design, usually lower in cost, more easily reconfigured and can be installed on twisted pair wiring. Baseband's major disadvantages are that it can support only digital data, not voice or video; if use were heavy, response time could be slow; and the cable can support only one LAN vendor.

A broadband system would offer many advantages. It could transmit voice and video as well as data. Various vendor systems could be accommodated on the same cable. Further, because of its higher data capacity, it could be used to transmit a baseband system (but not vice versa).

The committee decided that a broadband LAN was the best solution. It would provide the greatest functional flexibility and expansion potential, allow the company to add or change vendors and, according to Phillips' calculations, in the configuration Clark Fibers wanted would cost about the same as baseband.

The consultant also concluded that a broadband LAN would be the best long-term solution for the campus communications needs. In late 1981, he briefed top management on his findings. The committee viewed this as an important step in building upper management support for this complex technological decision, which would affect so many areas of the company. Once the committee decided to install a broadband LAN, two important questions remained: how much cable should be installed, and what broadband vendor should be chosen?

FORMATION OF THE MANAGEMENT COMMITTEE

While the working committee was struggling to meet the February 1982 deadline, the scope of the project increased dramatically. When Phillips first priced the pilot project, it would have cost approximately $50,000; as other departments became involved, the cost increased commensurately. This began to slow the decision process, because the committee was being asked to make decisions that would require resources beyond its control. It was having trouble deciding which vendor to choose, and there was even some doubt about whether the purchase should be deferred until a CBX had been developed that could answer the company's needs.

Unable to resolve these issues, the working committee requested that a committee of more senior managers be convened. This management committee was made up of members from the same departments as the working committee, but two levels above them. The committees met separately and reported to each other through memos. This procedure caused some frustration, since it was not always clear what had been decided and what remained to be done. The scope of the project and its cost, however, necessitated the involvement of the higher level of management.

In thinking about the project the committee considered the following issues.

Up to 80% of the cost of laying cable was in the labor to install it. Consequently the committees wanted to be sure that their choice afforded the greatest possible flexibility. They began by assuming they needed at least two cables, one for the broadband LAN and one for back-up. Clark Fibers was also considering buying a number of Wang office systems, and the committees wanted to keep open the option to install a WangNet, a two-cable broadband LAN system. Because the entire communications capacity is managed on the Wang system, it would be impossible to put any other vendor's system on the WangNet cables. This

meant Clark Fibers would have to lay at least four cables: one broadband, one back-up, and two WangNets. Then, after considering possible manufacturing applications, the committees decided to add a fifth cable. It would be needed to support CAD/CAM, which they felt was an important application to develop during the next decade, and the incremental cost would be only $7,000.

VENDOR SELECTION

During its deliberations the working committee had been looking at different vendors and their products. Steve Phillips sent vendors a questionnaire requesting more specific product information. The field was ultimately narrowed to four vendors: Alpha Systems, Pollux Data, Applied Communications, and Lydex. (Exhibit 4 gives statistics on each and on a representative baseband vendor, Centranet.) These vendors gave presentations to the working committee and provided references from users of their systems. After talking with all of the references, Murton and Phillips decided the Lydex system was worth investigating more closely. Steve explained:

We looked at the different vendors and talked with their users. Some of the users were very happy, others had mixed reviews, and in a short time we were down to a choice between the Alpha system and the Lydex system. The thing that impressed me the most about the Lydex system was that they did not exaggerate a single claim. That clarity impressed us.

THE RECOMMENDATIONS

In February of 1982, the committees recommended that the Clark Research Laboratory be wired completely with a Lydex broadband LAN, and that five cables be laid between the buildings on campus as a "backbone" to the system. Because the technology was so new, especially to Clark Fibers, the committee felt that three questions needed to be addressed before their recommendation was implemented. First, would the Lydex system work as well at Clark Fibers as it had at the user sites the committee had visited? Second, what would be the costs and benefits of the system? And third, how could the risk inherent in such a large project, involving new technology, be reduced?

To get actual experience with the system, and to become familiar with the Lydex system, R&D systems bought a test kit for $10,000, in June 1982. The test kit contained a small but complete broadband communications network. (See Exhibit 5 for a list of the test kit contents.) Steve Phillips designed three test situations to simulate the operating environment for the LAN at Clark Fibers: a small computer environment, a laboratory environment, and a mainframe environment. Each

was set up in turn and run for several weeks.

In the first two tests, Phillips performed such functions as file transfers, data retrieval, and remote printing, all without any difficulties. In the mainframe environment, however, some substantial problems arose. When terminals interacted with the IBM mainframe, they would display half a screen of data and then "go crazy." It appeared that the IBM was sending data to the network faster than the other connected equipment could receive it. Phillips called together technicians from IBM and Lydex, and together they worked out a solution that allowed the network to interface with the mainframe, but at a reduced data rate. This solution was important to Clark Fibers because communication with the mainframe was considered a

	Centranet	Alpha Systems	Pollux Data	Applied Communications Net/One (Broadband)	Lydex
Services					
Data	All data	All data application	All data	All data	All data
Voice	No voice	Teleconferencing/ store forward	Voice		No voice
Video	No video	Digitized voice, video	Video	Video	Video
Topology	Bus	Tree	Tree	Tree	Tree
Access method	CSMA/CD	Some dedicated channels, some reservation (switched)	Circuit switched and some fixed, point to point	CSMA, CSMA/CD FDM	CSMA/CD, FDM
Transmission media	Baseband coaxial	Broadband coaxial	Broadband coaxial	Broadband coaxial	Broadband coaxial
Maximum distance	1–2M (approx.)	50Mi	50Mi	8–9 Mi. approx.	35Mi approx.
Max. # of nodes	1024	4,090	Approx. 15,000	1,500	25,000
Max. # end users	1020	16,360	60,000	36,000	65,000
Total throughput		14Mbps	5Mbps	25Mbps	10Mbps
Max. trans. rate	10Mbps	56Kbps	5Mbps	2Mbps	
Trans. modes	Digital	Analog/digital	Analog/digital	Analog/digital	Analog/digital
# current users	120	Over 100	Approx. 300	New Product	150
Pricing					
Avg. price end-user device connected	$500–$1,000	$1,300	$500–$1,000	$500	Approx. $600
Price includes	Cables, transmission and connection, hardware and software, applications support, tech. support, documentation, training, troubleshooting	Net management and connection, hardware; net operating software documentation; tech. support	—	Transmission, connection and network management, hardware, network operation and applications, support software, installation, documentation training	Net hardware, operation software applications, support, installation, technical support documentation training, troubleshooting, with office connections to X.25

EXHIBIT 4

QUANTITY	DESCRIPTION
	Lydex broadband LAN test kit equipment
1	Head-end Frequency Translator
3	LAN MODEMS
	(two port. asynchronous radio frequency modem)
1	LAN MODEMS
	(eight port. asynchronous radio frequency modem)
	Small Computer Environment Equipment
2	DEC VT100 terminals
1	DEC 11/23 minicomputer
1	IBM Personal Computer
1	Tektronix 4052 graphics terminal
1	Epson printer
1	u-Mac data acquisition system
	Lab Environment equipment
1	DEC DP 11/44 computer
3	VT100 terminal
1	Tektronix 4014 graphics terminal
2	VISTA chromatographic data acquisition system
1	BOD data acquisition system
1	DEC LA 34 printer
	Mainframe equipment
1	VT100 terminal
	Front-end processor
2	Tektronix 4014 Graphics terminals
1	Lear Siegler CRT (ADM-3A)

EXHIBIT 5   Test kit contents.

crucial service on the LAN. Although it took eight months to iron out this difficulty, the system eventually passed all its tests.

Phillips and others in his department then put together an estimate of the economic impact of the LAN. To assess the magnitude of potential savings, Phillips looked at the current and projected communications needs and compared the cost of meeting them in the current way with the cost of providing the same service through the Lydex system.

There were five categories of savings: R&D personnel productivity improvements; MID personnel productivity improvements; MID development and support service no longer required; reductions in expenditures on computer resources (capital costs or hardware/software); and new capabilities. Appendix A provides an abbreviated version of Phillips' cost justification paper.

In R&D, the important savings would come from reduced set up time and better use of professionals.

MID personnel productivity would increase because the LAN would allow R&D people to transfer files and get data from remote locations. Previously these files were supplied by MID personnel on floppy disks.

The need for MID development and support services would decrease because MID would have to spend less time interconnecting R&D computers.

Reductions in computer resources would come from the fact that the LAN would allow more sharing of hardware and software.

Although Phillips felt unable to put a dollar amount on the savings from new capabilities, their impact on the organization seemed likely to be substantial.

While most of the savings were expressed in

terms of man years saved, there was no intention of laying off R&D personnel. Instead, the committees argued that the LAN would free up professionals and other personnel to perform other tasks more often and more efficiently. Phillips commented:

> The real savings with the local area net will not come from staff reduction. Instead, it would allow $40,000/year professionals to do something more productive with their time than move terminals around and hook up wires. Furthermore, the quality of the decisions will be better because the scientists will have better access to data and data analysis techniques. For example, they might see that they were heading down a dead end in an experiment in one or two weeks instead of four or five.

Commenting on the potential savings of the system, Phillips continued:

> The first cost justification I put together, based on our results of the Lydex test kit, had a pay-back of five months and an IRR of 235%. When I showed these numbers to my managers they said it looked over-

optimistic. So we cut it down some. I divided some of the projections by 3 and it still looked too good to be true, so I divided it by 2 again. Even so the project showed a 43% IRR, based on an investment of $500,000 and a productivity improvement of $184,000 for the first year and $394,000 per year thereafter.

THE DECISION

Despite this persuasive argument, Peter Lowell felt uneasy. It was crucial that the company support a robust and productive R&D effort, but the projected savings were soft. He wanted to feel sure that the committees had picked the best option. The short-term plan looked clear enough, but who would ultimately be in control of the system? Furthermore, 1982 had been a tough year, and he wondered if this were the right time to invest in a new technology, especially when it promised such uncertain results. Other fiber companies, however, were already investing in their R&D departments to improve productivity. If Lowell waited, it might be too late.

APPENDIX A   Clark Fibers, Inc.
Description of Benefits

The benefits to be derived from the LAN can be categorized by:

- R&D productivity improvements (1 man-year prof. = $33,750 and 1 man-year tech. = $18,750).
- MID personnel productivity improvements (1 man-year = $45,000).
- MID development and support services no longer required (1 man-year = $45,000).
- Reduction in computer resources (capital cost hardware and software).
- New capabilities.

The benefits of the LAN are difficult to quantify since we have limited experience in the use of LANs in R&D. NO DOLLAR SAVINGS HAVE BEEN ATTACHED TO THE NEW CAPABILITIES CATEGORY. As an example the first benefits are described and the yearly savings listed. (See Table 1.)

LABORATORY AUTOMATION

The LAN will reduce the time it takes researchers to set up experiments that involve automation. Instead of having to carry a terminal to the lab, connect it to a computer port, start the data acquisition program and then connect the instrument, researchers can do these functions from any terminal attached to the LAN or remotely through a gateway. In the R&D phase of the LAN project, it was shown that 50%–70% of the set-up time could be saved. The time saved in set up should allow researchers to complete more experiments.

- R&D personnel productivity improvement—$45,000 (2.7 man-years tech. or 1.25 prof.).

- MID personnel productivity improvement—$11,250 (1/4 man-year prof.).
- New capability: Researchers will be able to easily connect a variety of instruments and terminals that previously were not connected, thereby increasing the amount of data that can be automatically acquired.

COMMUNICATION BETWEEN NONCOMPATIBLE DEVICES

Most instrumentation manufacturers use their own protocol for communicating with other devices, causing problems when data need to be transferred to computers.

The Lydex modem has several features which allow incompatible equipment to communicate. The features most useful are the baud, stop bit, echo and parity translation capabilities. The modem also allows several types of flow control. During the LAN R&D phase, these features were used to connect a data acquisition device to a DEC PDP-11/23 via the LAN in about 15 minutes. To write and test the software necessary to provide communications between the data acquisition device and the 11/23 took about one week without the LAN.

- MID development and support services no longer required—$11,250 (1/4 man-year prof.).
- R&D personnel productivity improvement—$18,750.
- New capability: Researchers will have a simple way of connecting research instruments to reactors, computers, etc. Some applications which were not considered in the past can be implemented.

($)	R&D Personal Productivity Improvement	MIS Personal Productivity Improvement	Elimination of Need for MIS Services	Computer Resources (Capital Cost) (Hardware and Software)	New Capabilities
Laboratory automation***	45,000	11,250			**
Communication between non-compatible devices***	11,875		11,250		**
Minicomputer port utilization	33,750	3,750		44,250	
Minicomputer communications and software	22,500	3,750		10,500	
Multifunctional terminals	16,875			26,250	**
Installation costs	33,750	3,750	11,250	21,750	**
Graphics	45,000				
Computer services	33,750				**
CAD/CAM	90,000				
Peripheral usage	11,250				
Subtotals	348,750	22,500*	22,500*	102,750	
TOTAL	496,500				

TABLE 1   Productivity impact and capital cost savings for Clark Fibers, Inc.*

  *Variable savings at 70% DCF.
 **No dollar values attached.
***Analysis of savings is attached as an example of how calculations were made.

GM MAPS THE FUTURE

General Motors began pushing for common standards for factory-floor communications when it found in 1980–81 that linking pieces of equipment together accounted for 50% of the cost of automation in its factories. As America's largest manufacturer, GM has the biggest problem hooking up production machinery. Of the 40,000 programmable devices that GM uses, from controllers to robots, only 15% can communicate with other types of equipment. But by 1990, GM expects to be using 200,000 programmable devices.

In late 1982, GM started a campaign to get suppliers and users of factory automation equipment to work on detailed, common standards for factory-floor communication, nicknamed MAP. Suppliers had to back MAP to keep their business with GM. GM will use MAP-standard communications in its showcase axle plant at Saginaw, Michigan, opening next year.

MAP is based on a broad, seven-layer framework for common communications developed by the International Standards Organization.

The seven layers are guidelines for defining common communications standards, ranging from individual bits of data to the context (e.g., transferring files, loading a programme onto a robot) of what is being transmitted. After two years of talks among companies in the MAP users' group, most of the details of how to apply the standard to factory communications have been worked out.

MAP standards will be applied first on a local "token-bus" network. In a token-bus network, one piece of equipment (e.g., a machine tool) can communicate with others only when it receives a token which circulates on the network. Possession of the token allows the machine tool's controls to deliver a message immediately, much like possession of the conch shell at a primitive meeting. GM and other manufacturers prefer the token-bus network to the Ethernet network which is gradually emerging as a standard in office communications. They claim it can pass on urgent messages faster. That matters more when networks control heavy machinery.

The Economist, September 14, 1985. © 1985 *The Economist*, distributed by Special Features.

9 STANDARDS, REGULATION, AND DIVESTITURE

STANDARDS

REGULATION

DIVESTITURE

AFTER READING THIS CHAPTER, YOU WILL UNDERSTAND THAT:

- Effective business telecommunications requires international agreement on standards.
- National and international groups are working on these standards.
- The United States is moving toward deregulation, as illustrated by the divestiture of AT&T.
- AT&T divested itself of 22 operating companies so it could enter new areas of business previously denied to it.

Business telecommunications is like a three-legged stool, supported by three different areas:

- the technologies, like fiber optic transmission
- the products available, like integrated services digital networks
- the regulatory environment, which influences both the technologies and the products available.

Much of this book is about the basic technologies. Since new products come onto the market every day in numbers that would stagger anyone trying to keep up-to-date, we have discussed products only when they help to explain the technologies. In this chapter, we will concern ourselves with the third area: the regulatory environment. The proportion of material in this book devoted to regulation might suggest that it is of minor concern to business telecommunications, but nothing could be further from the truth. Comprehension of the regulatory environment is essential for anyone who wants to understand business telecommunications.

The importance of regulation (as well as *deregulation*) has to do with a free-market economy; if there were no governmental regulation at all, supply and demand would decide which products and services are purchased, which technologies are developed, and where research effort is applied. But in a completely free-market economy, the multitude of competing systems and technologies would all but make business telecommunications impossible; that's why *standards* are a necessary corollary to regulation.

STANDARDS

If you move a lamp from your home to your office, you'd expect to be able to plug it into a socket and turn it on. When you drive on an interstate highway, you don't have to drive on the right in some states and on the left in others. General agreement on conventions and standards makes life simpler.

So it is with business telecommunications. Telecommunications standards emerge in various ways: several manufacturers get together, or consumers apply pressure, or one supplier (e.g., IBM) is sufficiently dominant in the market. In some cases, countries get together to agree on standards: the most prominent organizations involved in setting telecommunications standards are shown in Figure 9–1. The three that operate at the international level are:

- **CCITT** (*Comité Consultaif Internationale de Télégraphique et Téléphonique*) is the standards organization of the International Telecommunications Union (a United Nations agency), and arbitrates

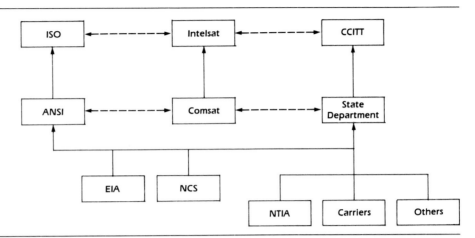

FIGURE 9-1 *Standards organizations.*

telephone and data communication standards among governments. The United States and other nations are voting members of CCITT, and corporations like AT&T are consulting members. CCITT has developed such standards as X.25 for terminals operating in the packet mode on public data networks (see Chapter 4).

- **ISO** (International Standards Organization) is a voluntary organization composed of the national standards committees of participating nations, including the American National Standards Institute. The ISO and CCITT coordinate their activities. The ISO is responsible for the OSI model, encryption conventions, and other standards.

- **Intelsat** (International Telecommunications Satellite Organization) was formed in 1964 to develop and promote the use of satellites. Over a hundred countries, including the United States, belong to Intelsat, which furnishes channels for telephone, data, television, and other satellite transmissions.

In the United States, the domestic standards organizations are:

- **ANSI** (American National Standards Institute) is a voluntary body and a member of the ISO.

- **EIA** (Electronic Industries Association) is a trade association that has helped to develop standards in North America such as the RS 232C interface.

- **NCS** (National Communications System) is the arm of the General Services Administration that is responsible for U.S. government standards. Federal telecommunication standards are developed jointly with the National Bureau of Standards.

- **Comsat** (Communications Satellite Corporation) is a private company created by Congress in 1962. Comsat represents the United States in Intelsat.
- **NTIA** (National Telecommunications and Information Administration) is part of the U.S. Department of Commerce, and helps develop federal telecommunications policy.

As this overview of the standards organizations suggests, it is generally agreed that it makes no sense to have competing telecommunications standards. This same philosophy is often the impetus behind regulation.

REGULATION

When the telephone industry was just getting started, the sensible policy was to avoid duplication of expensive facilities. Accordingly, the U.S. government awarded private monopolies to individual companies and made them subject to governmental regulation. These companies were called *common carriers*, and they provided services to all at the same prices and without discrimination. The Interstate Commerce Commission (ICC) had jurisdiction over these carriers because telegraph lines were located on railroad rights-of-way.

The Communications Act of 1934 made the regulation of interstate radio, telegraph, and telephone communications the responsibility of the *Federal Communications Commission* (FCC). Public utilities in each state were given the responsibility for local communications. Some FCC regulatory milestones include:

- **Specialized common carriers.** In 1963, Microwave Communications, Inc. (MCI) asked the FCC for approval to offer microwave transmission services between Chicago and St. Louis. The FCC gave its approval, and created a new industry which provided specialized services in contrast to the broad, basic services of the common carriers.
- **The Carterfone decision.** Until 1969, customers could not attach any "foreign" devices (devices not made by a common carrier) to the public network. Common carriers argued that their responsibility for end-to-end service required authority over *all* network facilities and equipment. The Carter Electronics Company, which marketed a device for integrating private radio communications into the public network, fought this restriction and won the right to connect their product to the telephone company equipment. This opened up the industry to non-AT&T equipment, and transformed a monopoly into a competitive market.
- **MCI.** Specialized common carriers further infiltrated AT&T's monopoly in 1975 when the FCC ruled that MCI's customers could use

the telephone company's local loops to dial into the MCI long-distance network. Other vendors like U.S. Sprint now offer similar long-distance services.

■ *Value added carriers.* A 1976 decision allowed value added carriers to lease facilities from common carriers or specialized common carriers, and to add special capabilities such as packet switching.

In addition, the FCC has conducted three inquiries to gather information, develop policies, and provide guidance for the computer industry.

■ *Computer Inquiry I* (1966–73) defined data processing and message switching, and all common carriers (except AT&T) were allowed to do both.

■ *Computer Inquiry II* (1976–81) permitted common carriers to split into separate subsidiaries offering *basic services* (e.g., telephone service) and *computer-enhanced communications products* (e.g., packet switching). Basic services would be regulated, enhanced products unregulated.

■ *Computer Inquiry III* (1985–86) signalled the FCC's adoption of the integrated services environment, and established a framework for competition and regulation of integrated information systems. This put the U.S. stamp of approval on such international standards and agreements as the OSI model.

The FCC (at least for the moment) is moving toward deregulation, as the following quote suggests:

The United States has never had a coherent national policy for the development and use of information technology. Such information policy as we have had has been closely tied to the evolution of particular technologies, each conceptually distinct from the others, and each with its own special uses, markets, and associated interplay of political and economic interests. As a result, the mail, the telegraph, the telephone, print publishing, broadcasting, cable television, automated data processing, and satellite communication, each in its own time and each in its own way, have given rise to a separate body of law and interpretation and in many cases to a separate structure of interpretative institutions. Each technology, moreover, has been approached from a somewhat different policy perspective. The conditions encouraging entry into the marketplace have varied greatly from one technology to another. Government approval has been required in some instances, while free entry, subject only to antitrust constraints, has elsewhere been the norm. Accordingly, competition has been forbidden, permitted, or required. First Amendment rights, privacy protection, copyright, and freedom of information requirements have all been affirmed differently in different contexts.

This pervasive fragmentation of policy initiative and responsibility could be sustained so long as there were valid technological grounds for distinguishing a letter from a phone conversation, a television image from a wired birthday greeting, and a telephone switchboard from a computer. Today, however, computer-based information technology is calling all such distinctions into question and, with them, many of the premises, perspectives, and expectations on which our nation's information technology policy, and particularly its regulatory policy, have heretofore been based.[1]

DIVESTITURE

One important result of this move toward deregulation was the divestiture of AT&T. In January 1982, the Justice Department and AT&T settled an antitrust suit. In the settlement, AT&T agreed to give up its 22 operating companies and, in return, was allowed to enter previously forbidden (and unregulated!) markets, including data processing, cable television, and information distribution.

Prior to divestiture, AT&T consisted of a Long Lines Division (responsible for long-distance telephone service), Bell Telephone Labs (research and new product development), Western Electric (manufacturing), AT&T International, and the 22 operating companies which provided 85% of the telephones in the United States with local service (see Figure 9–2, top).

By divesting itself of the 22 operating companies, AT&T in effect traded its secure position as the main supplier of basic telephone service for an opportunity to compete in new, mostly unregulated markets (long-distance service is still regulated by the FCC). A new unregulated subsidiary, AT&T Information Systems (see Figure 9–2, bottom) now manufactures and markets computer hardware and software and supplies advanced communication networks.

The 22 operating companies continue to offer local telephone service, which is still regulated by the public utilities in each state. These 22 companies, called Bell Operating Companies, are grouped into seven regional holding companies.

To sum it all up, AT&T no longer has control over the operating companies. But AT&T is now free to enter new areas of telecommunications, and is quickly becoming a vigorous competitor in the communications hardware and software markets. AT&T can, for example, go after the markets for local area networks, word processors, and personal computers.

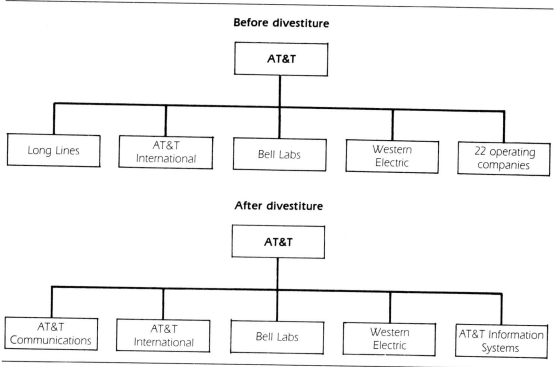

FIGURE 9–2 AT&T before and after divestiture.

Standards are critical in telecommunications. Office and factory efficiency requires that we agree on network standards that allow equipment from any vendor to exchange information freely with any other hardware on the network. The article at the end of this chapter, "DEC Moves to Support International Standard Protocols," provides some evidence that we are moving in this direction.

Regulation is also a critical factor in business telecommunications. It can shape a market, forcing supply and demand to conform to regulatory forces rather than to the free market, and affect the technologies that are offered and the products that are manufactured.

THE BRISFIELD COMPANY

What does the Brisfield Company's business telecommunications experiences (see the case study at the end of Chapter 1) suggest about standards? Domestically? Internationally?

SUMMARY

1. Effective business telecommunications requires international agreement on standards.

2. National and international groups are working on these standards.

3. The United States is moving toward deregulation of the telecommunications industry.

4. Deregulation is a logical step since there are no longer any technological grounds to distinguish voice from data, text, or image.

5. One example of deregulation is AT&T's divestiture of 22 operating companies so it could enter previously forbidden markets.

6. The regulatory environment can modify markets, so it must be understood by the individuals who operate in those markets.

REVIEW QUESTIONS

1. What do CCITT, ISO, and Intelsat do?

2. How about ANSI, NCS, Comsat, and NTIA?

3. Why was the ICC given jurisdiction over the telephone industry?

4. What was the importance of the Communications Act of 1934?

5. Review the FCC regulatory milestones.

6. Discuss the three computer inquiries.

7. Defend regulation. Now defend deregulation.

8. Is the FCC moving toward regulation or deregulation?

9. Describe the divestiture of AT&T. What do you think about it?

10. Standards are critical in telecommunications. Why?

11. Comment on the importance of the regulatory environment to the telecommunications market.

12. Read the article beginning on page 169. Why would DEC commit to sweeping, expensive software modifications to conform with international standards?

ENDNOTE

1. The Steering Committee on Computer-Based Information Technology and Public Policy of the National Research Council, *Computers, Communications and Public Policy* (Washington, D.C.: National Academic Press, 1981), report of a workshop at Woods Hole, MA, August 14–18, 1978, 26–27.

DEC MOVES TO SUPPORT INTER-NATIONAL STANDARD PROTOCOLS

JOHN DIX

MAYNARD, Mass.—Digital Equipment Corp. said last week that it will modify its network architecture during the next three years to comply with international standard protocols, a commitment that the company said is aimed at facilitating communications between different makes of computers.

In a sweeping announcement, DEC said it will do the following:

- Adopt the approved levels of the International Standards Organization's Open Systems Interconnect (OSI) network reference model.
- Build a software product implementing the OSI's transport level.
- Support the CCITT X.400 messaging recommendations.
- Modify its current X.25 support to conform to the revised CCITT standard.
- Provide hardware and software support of General Motors Corp.'s Manufacturing Automation Protocol (MAP).

Users contacted for comment on the DEC announcement said they were interested and pleased with the commitment but that the real importance of OSI support will only be realized if other vendors follow DEC's lead.

Of perhaps the most significance is DEC's intention to use the four lower layers of the seven-layer OSI network reference model in its Digital Network Architecture, industry analysts said. Computer systems of various makes that comply with OSI-specified protocols will be able to communicate, or so is the intent.

"OSI is going to provide users with a great deal of vendor flexibility by enabling information systems from various manufacturers to communicate. Users won't be tied into one manufacturer's implementation," said Harold Foltz, executive director of Omnicom, Inc., a consulting services company in Vienna, Va., that works with computer vendors to help apply these standards. "This is a key milestone in the evolution of OSI."

DEC will, during the next three years, adopt OSI layers one through four, the only OSI layers that have been ratified as international standards, the highest level of specification, according to John Adams, manager of strategic planning and marketing for DEC's Distributed Systems group.

Level 4 of the model, the so-called transport layer, provides for virtual circuits between computers and ensures that data arrives correctly and in the order sent, Adams explained. "It is the foundation for other network capabilities."

Even though OSI layers one through four are now international standards, OSI is still a reference model and, as such, offers a menu of class options. Vendors that implement different OSI classes will not be able to communicate, Adams said.

For this reason, DEC will try to build a superset transport that incorporates components specified by user-driven standards, specifically GM's MAP and Boeing Computer Services Co.'s Technical and Office Protocol (TOP). MAP and TOP are modeled after OSI and use a similar transport layer but differ in their specification of layers above and below the fourth layer.

Adams said that for upper levels of the OSI, DEC will replace its protocols where practical and coexist with the standard where needed. "For DEC-to-DEC communications at the applications level, we will probably maintain our own proprietary protocols because it will allow us to support more functions and achieve higher performance than we would if we supported the industry-standard protocol," Adams said.

"DEC's announcement shows the maturity of OSI," noted John Heafner, chief of systems network architecture with the National Bureau of Standards in Gaithersburg, Md. "Large organizations have a need to buy from multiple vendors, and in order to put multivendor products together, they have to have some kind of industry-supported standard."

Some analysts, however, fault OSI because of its options. "While OSI appears to be a standard, it is implemented slightly differently by almost everybody," said John King of James Martin Associates, a consultancy in Carmel, Calif. "OSI really isn't a standard; it is, as they call it, a reference model."

The first product that DEC will build to OSI will be the VMS OSI Transport Service. This product will reportedly facilitate task-to-task communications between a program running under VAX/VMS and a program running on another vendor's system that implements the OSI model. The product is scheduled to be released in early 1986.

Other standards DEC has committed to include the latest release of the CCITT's X.25 and X.400 recommendations. DEC will modify its current X.25 network support to comply with the newly revised 1984 X.25 standard, the company reported. The latest revision specifies use of a symmetrical protocol that enables data terminal devices to communicate without going through a public data network, Adams explained.

DEC said that, in 1986, it would begin to incorporate the X.400 mail and messaging recommendations into its products. This standard reportedly specifies how OSI-compliant networks exchange electronic messages.

Finally, DEC also committed to developing and delivering hardware and software systems that comply to Version 2.1 of GM's MAP. In 1986 DEC will deliver software for VAX/VMS and Microvax/MicroVMS, the company said. Controllers and modems that implement MAP protocols will also be made available at that time.

10 TELECOMMUNICATIONS AND MANAGEMENT

OFFICES
THE INTEGRATED NETWORK
THE COMPETITION
THE MARKETPLACE

COMPUTER-INTEGRATED MANUFACTURING
Implementation
Factory Networks

AFTER READING THIS CHAPTER, YOU WILL UNDERSTAND THAT:

- The productivity of white-collar workers is at an all-time low.
- Technology, especially telecommunications, can help improve office productivity (e.g., cut costs, increase output, and improve the efficiency of information flow).
- Smart businesses use information technologies as a competitive weapon (e.g., to raise revenues, improve marketing, and increase profits).
- Managerial concerns (e.g., costs, quality, flexibility) also apply to factories.
- Business telecommunications is a complex mixture of hardware, software, and firmware.
- Computer-integrated manufacturing helps factories survive.

OFFICES

In the nineteenth century, a majority of Americans worked on farms, but as scientific advances were translated into applied technology, our economy changed and manpower moved from agriculture to manufacturing. In this century, we have seen jobs migrating from the factories to service industries: those businesses involved in the handling, processing, and distribution of goods and services. If the computer had not been available to handle this sudden proliferation of clerical tasks and paperwork, we would have had to invent it.[1] Manual check processing alone would be so expensive without computers, few would be able to afford checking accounts.

Today, some 58 million workers—over half of the work force in the U.S.—have white-collar office jobs. According to the American Productivity Center, this group can be broken down as follows:

- 35% clerical
- 31% professional and technical
- 22% managerial and administrative
- 12% sales

Together, these workers account for some 70% of the total payroll.

But while the number of white-collar workers has grown, their output has dropped.[2] In fact, studies show that white-collar workers are generally only 40 to 60% productive.[3] Of course, no one is 100% efficient, but a 5% increase in efficiency would have a major impact on the profitability of a corporation. The white-collar worker is not the problem, however; it is the way that he or she uses his or her time that is inefficient.

One study suggests that these white-collar workers spend their time at work as follows:[4]

- 8% reading
- 13% creating documents
- 46% meetings (including phone calls)
- 8% analyzing
- 25% less productive activities

Some of these less productive activities can be eliminated through office automation:

ACTIVITY	SOLUTION
seeking information	data base management systems
placing calls	electronic mail
corrections and revisions to reports	word processing
attending meetings	teleconferencing

The United States' real business output grew by 18% between 1978 and 1985, and the number of white-collar workers rose by 21% during the same period. The problem is that white-collar productivity *dropped* by about 10% during these years. Faced with higher payroll expenses and less productivity, many organizations will have to cut white-collar workers just as savagely as they cut blue-collar workers during this same period.[5]

——————— THE INTEGRATED NETWORK ———————————

It's clear that office productivity has to improve in America, and the first step is to manage the office. The office can be managed and made more efficient if you define information to be its product. With this orientation, traditional management tools can be applied, such as cost/benefit analyses, goals based on requirements of the product, align organizational structures to produce the product, and base capital investment decisions on return on investment.

The second step is to recognize the importance of information as a corporate resource. Even though corporate and individual success depends on how well information is used, few managers take full advantage of this fact.[6]

Two technological trends have also emerged: first, distinctions among specific technologies (e.g., data processing and word processing) have blurred; and second, previously unrelated technologies are being interconnected through the use of integrated telecommunications networks. These two trends are leading us toward one comprehensive information technology: the integrated network. Users of the integrated network will access information in the form of data on computers, text on word processors, images on microfilm, and voice on the network itself. The network will also improve the productivity of managers and professionals by providing services such as the following:

- electronic mail
- voice messaging
- access to data bases (internal and external)
- access to text
- information manipulation
- decision support systems
- graphics
- access to microfilm
- local administrative support
- computer conferencing
- teleconferencing

That is, by automating the information resources of managers and professionals, we can raise the quality of their decisions, cut costs, and improve productivity. The appendix to this chapter provides some additional background.

——— THE COMPETITION ———————————————

For most organizations, it's all they can do just to keep up with the changes in information technologies, but a few companies, such as American Hospital Supply Corp. and American Airlines, have been investing in these technologies (and the skills needed to use them) for years. Many of these farsighted firms have reached the point where computers and telecommunications are used for everything from product development to market share. For example:

- to cut sales expenses, "cold" leads are qualified by telephone, and then computer programs are used to rank the prospects
- customers can access the supplier's data base directly, which lets customers track their own orders
- toll-free numbers allow customers to suggest ways to improve products and services
- salespersons enter orders directly on portable computers, which speeds delivery and cuts paperwork.

Businesses that learn how to use information technologies strategically may actually increase revenues—not just cut costs.[7] For example, Merrill Lynch & Co. combined checking, savings, credit card, and securities account information on one computerized monthly statement, which automatically invested idle funds in an interest-bearing money market account. This Cash Management Account was one of Merrill Lynch's most successful new products.

Our economy relies increasingly on information technologies, but computers and telecommunications are no panacea.[8] Automation means changing the way work is done, and that takes time and effort. U.S. companies have already spent hundreds of billions of dollars on computer-related solutions, but white-collar productivity has yet to reach its full potential.

Of course, any discussion of white-collar productivity must not obscure the problems of blue-collar productivity, which can be equally serious. Whatever is true for managers tends to be true in the manufacturing plant as well. When process engineers, data processing, customer service, quality assurance, and others are linked through networking, the result is often an improvement in quality, responsiveness, and productivity. Instead of isolated islands of technology, there is a plantwide exchange of information: factory floor and office are unified at last.

THE MARKETPLACE

Imagine you are a foreign manufacturer. How can you make a place for yourself in the U.S. market?

First, identify a high-volume product (i.e., a product that sells a lot of units), one with limited options and a firm market so sales will be predictable. Make certain its production processes are well known and readily available.

Now, examine the production system used by other manufacturers. You won't find one big problem; rather, you may find weaknesses such as obsolete machinery, indifference to product quality, increasing costs passed on to consumers as higher prices and concentration on short-term profits instead of long-term market position.

Next, create a production system that eliminates these inefficiencies. Using existing technology, upgrade the entire production process and cut out all the "fat." Now you're ready to ease yourself into the U.S. market.

This process is illustrated by the Fremont, California, plant that GM now runs as a joint venture with Toyota. When GM ran the plant, the 5000 workers assembled some 240,000 cars a year. Absenteeism was around 20%, and wildcat strikes were common. Under Japanese managers, about half the number of workers now use the same dated technology to build the same number of cars per year. The workers, the building, and the technology are the same; only the production system has changed.

Many U.S. companies—manufacturers of automobiles, airplanes, steel, and electronics—have learned this same lesson: the marketplace always rewards efficient manufacturing methods. We cannot maintain our standard of living without maintaining our industrial base as well. In today's market, delivery times are shorter, market conditions change faster, product modifications are more common, and customers demand greater responsiveness, so traditional competitive solutions are no longer enough.

COMPUTER-INTEGRATED MANUFACTURING

U.S. manufacturers have two weapons. First, change the way they run their organizations. Japanese businesses have fewer management layers, and decision making takes place at the lowest level possible. This kind of participative management would benefit American firms as well.

The second weapon is technology. *Computer-integrated manufacturing* (CIM) helps factories survive: computer-aided design lets the engineers

This section draws heavily on *Management Guide for CIM* (Dearborn, Mich.: Society of Manufacturing Engineers, 1986), 97; "Making the Leap into the Factory of Tomorrow," *Industry Week*, May 26, 1986, 41; and "High Tech to the Rescue," *Business Week*, June 16, 1986, 100.

design the product on a video screen; computer-aided engineering analyzes the product for performance and ease of production; and computer-aided manufacturing automates the shop floor machinery. The result is faster, cheaper, and better production.

Computer-integrated manufacturing also reduces the costs of direct and indirect labor, as well as some "fixed" costs like overhead. Robots don't make careless mistakes, so quality control is cheaper. Robots are faster, so lead time can be slashed. CIM factories can break even operating at 30% of capacity, compared to 65% in some conventional plants.

Since the objective of computer-integrated manufacturing is to achieve greater efficiency throughout the entire business—across the whole cycle of product design, production, and marketing—CIM is more than just a new technology. It is a business strategy which uses new and existing computer applications to share the combined information assets of the company.

Most firms have unconnected "islands" of automation: manufacturing automated protocol, office automation (such as word processing and data bases) computerized market forecasting, general ledger systems, computer-aided design, and many others. A huge amount of manpower and paperwork goes into keeping these islands from sinking. CIM links these islands digitally, essentially creating a single, shared data base which contains all the plans, routines, and data necessary to operate the whole company: factory, office, and everything else. The chief implementation problem is administrative, since CIM crosses all functional lines.

CIM is not a new concept; what *is* new is that the technologies are now cheap and powerful enough to make this strategy a viable one. In the last few years, all the elements have finally come together: hardware, software, communications, costs, and attitudes. In fact, the pressures of the marketplace may soon put a business that *doesn't* use CIM at a big disadvantage.

Implementation

As shown in Figure 10–1, the implementation of CIM by a manufacturing company occurs in four stages:

1. introduction of the CIM business strategy
2. recruitment of the necessary expertise
3. development of a specific CIM program
4. implementation of CIM.

It takes from four to six years to implement CIM, and there are many different ways to do it: a firm producing high-priced, customized items will not follow the same plan as a company manufacturing low-priced, high-volume items.

FIGURE 10–1

Management cycle for CIM.

1. Introduction
concepts of CIM business strategy
managers' roles
creating interest in CIM
developing the climate for CIM

2. Preparation
forming a study team for CIM
CIM opportunity candidates
conceptualizing CIM operations

3. Program Plan
proposal development
selling the proposal
program commitment

4. Implementation
managing the implementation
measuring and evaluating results
moving on to the next opportunity
sharing experience and expertise

Management Guide for CIM (SME, 1986), 11.

Factory Networks

The network is the heart of CIM. Equipment is tied together and information is shared via the network, which transfers files, loads new programs into robots, and so on. For example, manufacturing automation protocol (MAP), a leading factory networking standard embraced by GM and others, can be used to interface a production and material control system on an IBM mainframe with a quality control and production scheduling system on a Honeywell computer. Furthermore, MAP can connect devices from different vendors, such as process controllers, personal computers, and computer-aided design and computer-aided manufacturing work stations. The network also integrates automation techniques in manufacturing plants.

To summarize, computer-integrated manufacturing helps factories survive. CIM improves product quality, productivity, and competitiveness by sharing the combined information assets of the company. There is a real need for these improvements: a revitalized manufacturing company provides stimulating, challenging jobs, and by becoming more competitive, such a company uses more services such as banking, communications, education, health care, and travel. These services in turn help manufacturers. Strengthening the manufacturing sector helps maintain the industrial base and a high standard of living.

SUMMARY

1. The United States has a service economy, where over half the work force has white-collar jobs.
2. Although white-collar workers receive 70% of the total payroll, they are only 40 to 60% productive.
3. The white-collar worker is not the problem; it is the nature of the work.
4. Office automation provides affordable solutions to the white-collar productivity problem.
5. Information technologies are being integrated into one comprehensive information technology: the integrated network.
6. A few companies have used information technologies to give them a competitive advantage in everything from product development to market share.
7. Business telecommunications is a complex mixture of hardware, software, and firmware.
8. Computer-integrated manufacturing helps factories survive.

REVIEW QUESTIONS

1. What is the white-collar productivity problem?
2. Is the white-collar worker to blame? Why (not)?
3. What is the solution?
4. Previously unrelated technologies are being integrated. Comment.
5. How can information technologies be used strategically?
6. Comment on the CIM technique for improving factory productivity.
7. What does MAP have to do with CIM?
8. How does one implement CIM?

ENDNOTES

1. Frank Greenwood and Erwin M. Danziger, *Computer Systems Analysis: Problems of Education, Selection and Training* (New York: American Management Association, Management Bulletin 90, 1967), 1.
2. *White-Collar Productivity: The National Challenge* (American Productivity Center, 1982), 37.
3. For an elaboration of these ideas, see Frank Greenwood and Mary M. Greenwood, "Productivity Brief," *Journal of Systems Management*, November 1984, 38.

4. Harvey L. Poppel, "Managerial/Professional Productivity," *Outlook*, Fall/Winter 1980, 29.

5. "Is American Business Being Managed to Death?," *The Economist*, December 13, 1986, 71.

6. Frank Greenwood and Mary M. Greenwood, "Survival and Prosperity in the 80s," *Journal of Systems Management*, November 1982, 6.

7. See "Information Power," *Business Week*, October 14, 1985, 108–114.

8. William Bower, "The Puny Payoff from Office Computers," *Fortune*, May 26, 1986, 20.

APPENDIX
Management

Business telecommunications seldom fails for technical reasons. When it does flop, it is because of poor management, and especially because of inadequate planning.

We must therefore have a definition of management.[1] Once this complex idea is defined, we can reflect on how networks affect management, and how you can plan a telecommunications system to improve your business results.

Humanity has learned much about management over the centuries: empires rose or fell, armies conquered or were vanquished, and commercial enterprises prospered or failed. Unfortunately, much of this experience was not documented.

About the time of World War I, a French mining engineer, Henri Fayol, published a book defining management. It reflected his experience in taking over an almost bankrupt steel manufacturing company and converting it into a prosperous, integrated firm. His book was translated into English[2] and worked its way to the United States. Americans took to his management approach and many books are now based on Fayol's concepts. Here is the definition of management common to these books:

Universal concepts in management are equally applicable to commercial, religious, and governmental organizations. Business management differs from the others mainly in that the profit realized on invested capital is, for many, an important measure of managerial ability.

Because of physical and mental limitations, people must cooperate to achieve their goals. The skill of management is basic to this cooperation. The job of management is to get things done through others. In other words, management is the accomplishment of desired objectives by establishing an environment favorable to efficient performance by people operating in organized groups. Therefore, the subject of management is people, and not simply technology, facilities, finance, or anything else.

The managerial job is similar at all levels of an organization and in all organizations, when the individual is acting as a manager. The managerial job differs from that of the nonmanager because the manager is responsible for synchronizing the efforts achieved by subordinates.

The manager achieves coordinated effort toward accomplishing the goals of the enterprise by planning, organizing, staffing, directing, and controlling the activities of others. The essence of management is coordinating the activities of people, and it is achieved by means of the five functions defined here.

Planning *is the function of selecting the objectives of the enterprise and the* *policies, programs, and procedures for achieving them.*

Organizing. *The organizational function of the manager involves determining and enumerating the activities required to achieve the objectives of the enterprise, grouping these activities, assigning such groups of activities to a department headed by a manager, and delegating authority to carry them out.*

Staffing *is the function comprising activities that are essential in manning, and in keeping manned, the positions provided for by the organization's structure.*

Directing *is the executive function embracing activities related to guiding and supervising subordinates.*

Controlling *is measuring and correcting the performance of subordinates in order to make sure that the objectives of the enterprise and the plans devised to attain them are accomplished.*

As a manager you are responsible for the performance turned in by your part of the organization. You must therefore have the authority to require that subordinates conform to decisions. Authority makes your position real. But it must be used very carefully, only as a last resort, in fact. Selling and persuading are the first choices. What you can do with formal authority is very limited.

Although this is a somewhat sketchy but reasonably usable definition of management, we must keep in mind that management will always be primarily an art, because the subject of management is people. There is no substitute for on-the-job managerial experience.

Individuals who understand that management is this distinct kind of work will, however, probably understand their job better and improve their performance.

Management is thus getting things done through other people. Its essence is coordination, which is achieved by planning, organizing, staffing, directing, and controlling the activities of people. Coordination is accomplished with information: information about what is planned, who is

supposed to do which jobs, how actual performance compares to budget, for example.

As information is an important corporate resource, your business telecommunications system can be regarded as your firm's nervous system. This system can help coordinate the parts of your company (people, money, and machinery) and synchronize your business with its markets. Because networks are information-handling devices, they can often help improve a business information system, contributing to better coordination so that you manage more effectively.

Andrew S. Grove, president of silicon valley's Intel Corporation, characterizes the relationship of information and management as follows:

> *It's obvious that your decision-making depends finally on how well you comprehend the facts and issues facing your business. This is why information gathering is so important in a manager's life. Other activities—conveying information, making decisions, and being a role model for your subordinates—are all governed by the base of information that you, the manager, have about the tasks, the issues, the needs, and the problems facing your organization. In short, information gathering is the basis of all other managerial work, which is why I choose to spend so much of my day doing it.[3]*

SYSTEMS ANALYSIS AND DESIGN

Another expression for planning is *systems analysis and design.*[4] Broadly defined, systems analysis and design is the function of devising an appropriate procedure to achieve a particular purpose. It is composed of an analytical component and of a creative design element. Good planners and efficient administrators have always done systems work. Alexander the Great and Julius Caesar are likely to have done systems analysis and design in this sense.

Systems analysis cannot begin until the purpose of the system has been defined. When a systems analyst is assigned a project, he or she is seldom—if ever—given a clear definition of the problem. For example:

- "The legal office's typing is backed up for two weeks." This may mean a better system is needed for word processing, or it could simply be evidence that the lawyers are maneuvering and building a case for more private secretaries.
- "The personnel department won't provide the required ERISA reports on time." Maybe a special software package is needed for the computer to process the reports efficiently, but perhaps the personnel department has been too slow providing data to data processing and this delayed input created late input.
- "The plant people can't make sense out of the huge inventory printout." Computer-output microfilm may be needed so the inventory records are more readily accessible, but the problem may be that the clerks have never understood how to read the computerized inventory records and all that is needed is a little instruction.

The analyst's first task is to define the problem. He or she must systematically inquire as to the information that is needed: by whom, when, in what form, in what way it would best be used. Existing systems may give good leads for finding the answers to these questions, but answers must be cross-checked against views expressed by all involved. The analyst must also observe the environment and circumstances in which the system will operate.

The result of this step is a statement of requirements for management's acceptance. Such a statement usually indicates:

- what information the system should supply
- what data are needed to create that information, including the sources of such data, and the circumstances in which they can be obtained
- the basic steps by which the required information can be derived from the data
- the practical limitations imposed by the environment.

Now that the problem has been defined, it can be analyzed. Begin by identifying all the parts of the problem and then considering each in turn, with particular emphasis on how each part relates

to the rest. It is usually necessary to determine the relative importance of the different parts as they relate to the overall purpose of the system. Some system problems are simple and can be solved with familiar approaches. Others may be very difficult and need novel solutions.

Where problems demand special solutions, there are two ways to proceed:

1. the analyst devises his or her own solution, which requires experimentation and cost analysis
2. the analyst inquires how others have solved similar problems, which requires research as well as cost analysis.

In any case, the objective is to reach a thorough and balanced understanding of the nature of the problem as a whole and of its constituent parts.

There are usually many possible systems that the analyst could devise to achieve a given purpose and they probably differ widely in suitability and cost. The analyst first makes sure that he or she understands all the factors in the problem, and then he or she allows sufficient time to reflect on the options open to him or her. The overriding consideration is that the proposed system be humanly, technologically, and economically feasible.

Most people are resistant to change, and a change of system is no exception. Participation and communication are helpful tools. Everyone who will work with the system is involved, listened to, and publicly credited for contributions. Forthright communication of what is being done, why, and how each will be affected is almost always the correct approach.

In due course, the system is designed and presented to management, showing that it meets the needs in a practical and economical way. Assuming approval is given to proceed, the system is implemented.

Usually, the most important factors in implementing the system deal with installing equipment and training people. Each step in implementation is normally listed and timetables are negotiated. Responsibilities for the steps must be established. The progress of implementation is controlled and coordinated by means of reports and meetings, during which achievements are compared to the plan.

After the new system is implemented, the analyst needs to audit it in order to learn if it is achieving its goals. This normally occurs only after a shakedown period, which allows everyone to learn the new ways before any assessment is made. (Ten guides for doing systems analysis are listed in Figure 1.)

In summary, a manager uses information systems to help coordinate the activities of people, and information processing technologies are often woven into these systems. To this, we add the con-cept of systems analysis. Systems analysis is the function of devising an appropriate procedure to achieve a particular purpose. Systems analysis is thus the usual technique for creating and modify-ing information systems and is the normal method for incorporating information processing technologies. These management and analysis con-cepts provide background to individuals respon-sible for planning and managing business telecommunications.

---

1. **Systems have three basic functions:**
   - getting coordinated action
   - remembering this action
   - reporting the results

   To illustrate: a payroll system achieves the action of paying people on schedule, it remembers the details of that action (e.g., pay rates, deductions), and it reports the results (e.g., W–2s to employees at income tax time). A college registra-tion system achieves the action of registering students into classes, it remembers that action (e.g., who is in which class), and it reports the results (e.g., class lists and grade reports).

2. **Systems improve coordination by gearing individual efforts to overall objectives.** This requires careful analysis of how the work should be done, including: skills, procedures, data, machines, forms, and policies. Give particular atten-tion to transfer points, where the responsibility of one person ends, and where that of another begins.

3. **Systems have natural cycles.** Find the pattern of the transactions: where it starts, where it ends, and the process points between. This defines the system to work on. Ignore department "walls" because they seldom coincide with systems boundaries.

4. **Emphasize the work objective.** Always stress what results from this cyclical pattern. This work objective has to be related to the rest of the organization's efforts, coordinating the work.

5. **People make a system work.** Clerical people can make a poor system work or a good one fail. Involve them in the systems analysis early, listen to them, and implement their suggestions.

6. **Balance the system.** Every system has to balance costs and benefits by considering speed, cost, and quality. An im-provement in one may degrade another, so balance is sought.

7. **Documentation is critical.** The main tool for managing a system project is documentation, which is a written record of system details such as: purpose, what each work station does, what the input data are, and who gets what output. Create and maintain this documentation, which is your chief management tool.

8. **Forms and procedures are important.** Forms carry the data in many systems. They should be designed and main-tained with this basic role in mind. Procedures, too, are essential, since they help coordinate by defining who does what and when.

9. **Minimize data handling.** Handling data is expensive (it is analogous to handling material), so avoid it.

10. **Strive for integrated data handling.** Try to record data at the point or origin in machine-readable form, "capturing" enough to serve all subsequent needs, and handle the data electronically from then on.

FIGURE 1 Systems analysis guides.

ENDNOTES

1. This material is from Greenwood, *Profitable Small Business Computing* (Boston: Little, Brown, 1982), 5–8.
2. Henri Fayol, *General and Industrial Management* (London: Sir Isaac Pit-man and Sons Ltd., 1949).
3. S. Grove, *High Output Management* (New York: Random House, 1983), 51.
4. This material is from Greenwood and Greenwood, *Information Resources in the Office of Tomorrow* (Cleveland: ASM, 1980), 15–18.

11 NETWORK DESIGN

USER DEMAND
MESSAGE TYPES
FACILITIES
    Channels
    Equipment

COST AND PERFORMANCE
NETWORK DESIGN
NETWORK MONITORING

AFTER READING THIS CHAPTER, YOU WILL UNDERSTAND THAT:

- Network design is based on what users need.
- Networking facilities follow logically from demand analysis.
- Both demand and facilities are subject to cost and performance considerations.
- Procedures to help diagnose and correct problems are needed once the network is implemented.

Network design is based on what users need, so understanding the dimensions of demand is a logical place to start. The networking facilities that the user requires follow logically from demand analysis, and both demand and facilities are subject to cost and performance considerations. Therefore, we will discuss demand, facilities, and performance objectives, in that order.

USER DEMAND

The essence of any business is determining what people want and then supplying it. An economist might say that "aggregate demand is the key to economic activity"; in other words, demand for goods and services drives the economy. If people don't buy new cars, the steel mills and tire factories that supply the automotive industry will be idle, and *their* suppliers will be idle as well.

This concept applies to networks. Networks exist to meet the demands of the people who use them; for example, some people need to access data bases, while others have to run accounting, sales, or production applications. Outside users may use the network to place orders, determine shipping dates, or check account status.

Accordingly, estimating user demand is critical in network design. The applications and data bases that are needed now—and those that will be needed in the future—must play a major role in determining the network's resources, including message types, traffic volumes and peaks, and which locations within the company need information at which times.

MESSAGE TYPES

There are several message types common to most networks:

- *Data base inquiry.* An operator at a terminal requests information from a data base, and the information is returned to the terminal in *real time*—fast enough for the operator to make a decision and request more information. Terminals may be local or remote, and data bases may be at one or more sites.

- *Creating new records.* The operator can create a new record in the data base in real time (which makes the new record available immediately) or in batch mode. *Batch mode* transmits a set of transactions together, so individual records do not get into the data base until the whole batch of records is processed.

- *Modification and deletion.* The operator can modify or delete a record at a terminal in real time (this is called an *update*).

- *Process control.* Plant processes (e.g., oil refining machinery) are monitored automatically, and the information needed to control these processes is transmitted from a central location.
- *Message switching.* Software running on the front-end processor routes messages among the devices attached to the network.

These messages all share a general format which tells the devices at each end what to do and how to do it. As discussed in Chapter 6, a handshaking sequence is used to initiate the transmission, and once this sequence is complete the equipment can recognize the data transmission format.

Figure 11–1 illustrates a generalized message format. The *outer envelope* (the communications header and trailer) is system-generated: the header promotes the handshaking sequence and may include a routing segment, and the trailer handles link control, which may include error detection, routing, and message terminating functions. The outer envelope thus moves the messages between specific locations via SDLC or some other protocol.

Inside the outer envelope is the user-generated part of the message. The user header can be entered by the operator or inserted automatically, and may include a password, user ID, sending device ID, data base/application location, destination ID, message sequence number, and message priority. The user text is the actual message or response. The user trailer (generated automatically) keeps the blocks in order when a message is broken down into smaller blocks, and helps to detect transmission errors.

——— FACILITIES ——————————————————

Demand and facilities are two sides of the same coin: demand determines what facilities are needed, and both are fundamental to network design.

FIGURE 11–1————————————————————————————
Generalized message/
response format.

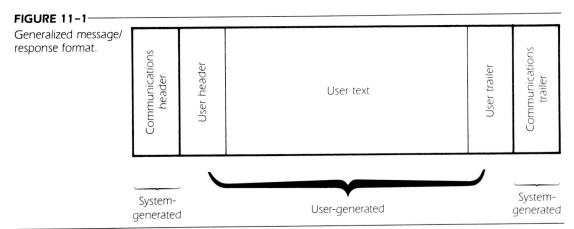

Communications header | User header | User text | User trailer | Communications trailer

System-generated | User-generated | System-generated

Facilities include software, transmission channels, and equipment, and since we discussed software in Chapter 5, this section will cover channels and equipment.

Channels

Transmission channels are paths that link the network's equipment, and may include wires, cables, fiber optic lines, and microwave and satellite relays. These links are usually leased from common carriers.

- *Local carriers.* The local telephone companies, including the 22 Bell Operating Companies, are the common carriers that connect your home and office to the telephone network via the local loop. Divestiture divided the Bell territories into 164 *local access and transport areas* (LATAs), plus 26 independent LATAs. Communications within each LATA belongs to the local carrier.

- *Long-distance carriers.* Communications between LATAs belongs to long-distance carriers, a group which includes AT&T, MCI, Western Union, and U.S. Sprint. Some carriers cover the entire United States, others serve only specific areas.

- *Value added carriers* not only provide a circuit over which users can transmit messages, but also provide some enhancements, such as packet switching. Tymnet, GTE Telenet, and AT&T Information Services all provide data switches that connect terminals to the appropriate host computer.

- *Satellite carriers* include Satellite Business Systems and American Satellite Corp.

- *Digital terminal systems* bypass the local loops and route satellite transmissions (both voice and data) directly from the user's own satellite antennas to the long-distance carrier. United Telecom is establishing digital termination systems in many U.S. cities.

- *Resale carriers.* Like wholesalers, resale carriers lease long-distance lines from carriers such as AT&T, and using their own switching equipment in many cities, they achieve economies of scale for small local subscribers.

- *Radio carriers.* Cellular radio carriers offer mobile radio communications. Markets are divided into cells, each with its own base station, and as mobile radio users move, their transmissions (both data and voice) pass from cell to cell. The FCC authorizes two carriers per market, one of which is the local telephone company.

- *International record carriers,* including AT&T, MCI, and RCA, provide international service from the United States.

Most of these carriers are regulated by two different agencies: the FCC has jurisdiction over telecommunications between states, and the public utility commission of each state has jurisdiction within that state. Each carrier submits a document called a *tariff* to the regulatory agency, defining the services they intend to offer, specifying the charges, and spelling out the obligations of the carrier and the user. Every service provided by every carrier is covered by a tariff (for instance, *FCC 260* covers interstate private line voice transmission services).

Common carrier services can be divided into three broad categories:

- *Message service,* which is what most households have. There is a fixed monthly fee plus a variable charge depending on the number of calls placed and their duration.
- *Bulk message service,* with a fixed monthly fee that covers a specified number of calls to a certain area. Extra calls are billed at a rate that is usually less expensive than message service rates.
- *Private line service,* with a fixed monthly fee for a dedicated, direct connection. The fee depends upon distance, not call volume.

Equipment

Most of the equipment used by networks employs microprocessors and is software controlled. The software that controls the devices must therefore be compatible with the software that controls the network, a key factor in deciding which equipment to buy. A typical network would include some or all of the following devices:

- *User terminals.* Dumb terminals simply accept inquiries and receive replies from the host computer. Smart terminals have their own limited intelligence and can do some local processing. Peripherals such as printers, plotters, and disk storage devices are often connected to terminals.
- *Batch terminals.* Remote job entry allows jobs to be submitted to a central computer through a batch terminal connected to the computer by a data link. Batch terminals often include tape drives, disk drives, and line printers.
- *Terminal controllers* (also called cluster controllers or line controllers) are for locations with more than one terminal. They control access to the network and distribute the host computer's replies so cheaper, less intelligent terminals can be used. Since terminal controllers need only one address, fewer ports are needed for the front-end processor.
- *Modems* are needed to transmit digital data over analog voice lines, as discussed in Chapter 2.

- *Multiplexers/concentrators* collect transactions from scattered sites and feed them into one, high-speed line to the central computer. We have already seen that multiplexers come in pairs, one at each end of the line. *Concentrators* have more intelligence, so only one is needed as long as the front-end processor is able to handle the transaction stream at the other end, directing the transactions to the correct terminals.
- *Front-end processors.* As discussed in Chapter 4, front-end processors handle line management tasks for the host computer.
- *Gateways.* When two or more networks that use different protocols are connected, a "black box" or *gateway* must be used to translate messages from one protocol to the other (see Figure 11–2).

COST AND PERFORMANCE

Network designers are concerned about the cost and performance of a telecommunications system. Costs include lines, hardware, and software; performance includes response time and throughput, as discussed below.

Costs

Line costs are set by tariffs, and are more-or-less fixed; in contrast, the price of hardware and software is often negotiable. Smarter hardware can sometimes reduce the number of lines needed (since the software provides the intelligence), so there may be cost benefits at a given performance level. Most organizations have firm policies concerning return on investment, and network cost estimates are often assessed in terms of when the network will pay for itself and what it will contribute to company earnings.

Performance

- *Response time* is the interval from when the last item was keyed in until the terminal begins displaying the response. Response time is thus the sum of data communications time (from the terminal to the host computer and back again) plus data processing time. Ideally, response time should not exceed two seconds; a performance objective might be to answer 95% of all inquiries within six seconds during peak network hours.

FIGURE 11–2 —————————————————————————
Gateway.

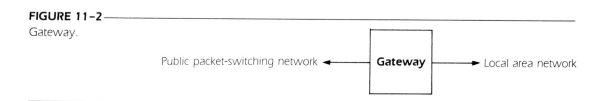

Public packet-switching network ◄——— **Gateway** ———► Local area network

■ *Throughput* is the amount of data that a channel can transmit in bits per second (bps); for example, a throughput of 2000 bps on a 2400-bps line. It is usually calculated for a one-way transmission.

There are performance trade-offs between response time and throughput. We can reduce response time by sending shorter messages (since all the bits can be checked quickly) or by using shorter message queues (which cut waiting time for message processing). But throughput is increased by longer messages (which provide more useful data with the same number of control characters) and longer queues (which smooth the traffic flow). In addition, support functions can raise response time (e.g., code translation, protocol conversion) and reduce throughput (e.g., adding application or data base identifiers to the message header).

————— NETWORK DESIGN ————————————————————

What the user needs is fundamental, so the volume of messages sent and received, as well as traffic patterns and their peaks, has to be estimated by site and by application. Network performance—response time, throughput, and support functions—should be projected to find the point where overloading or obsolescence will occur. These and other network design questions are listed in Figure 11–3.

Network design requires coordination of both internal and external elements: Internally, the designer must consider both the user's and the company's management, and externally, there are carriers supplying services and vendors selling hardware and software. All these people have their own interests and agendas, and a consensus between them will not occur spontaneously.

The implementation of a telecommunications network often follows these five stages:[1]

1. every part of the organization leases or buys whatever equipment and lines it pleases
2. the telephone system is centralized
3. separate data processing networks are optimized as a single network
4. voice and data networks are combined
5. all communications, including mail, travel, voice, and data, are integrated into a single network.

Different organizations will be found at different stages of network design. In the pages that follow, we will consider a company that has reached the third stage: design and implementation of a simple, on-line data system (see Figure 11–4).

FIGURE 11–3——

Network design
questions.

1. **What applications will be accessed over the network?**
   current applications, including type and volume of transmissions
   planned applications, including type and volume of transmissions

2. **What user population will the network serve?**
   distributed, where computer power and data bases are dispersed among interconnected sites
   centralized, where the processing is centralized and all terminals are connected to
   the central computer

3. **What network performance levels are desired?**
   response time, the interval between the user's request and the system's reply
   throughput, the amount of useful information passing through the system per
   unit of time
   error rate, the transmission errors per million characters sent

4. **Is location or characteristics of equipment more important?**
   location of equipment, such as terminals, modems, multiplexers, switching
   equipment, computers
   equipment characteristics, such as capacity, speed, number of lines served, full-
   duplex or half-duplex

5. **How will the equipment be interconnected?**
   private lines
   common carriers

6. **What are the costs and benefits of such a network?**
   costs, including hardware, software, and transmission channels
   benefits, including increased capacity and greater reliability

7. **What is the network implementation schedule?**
   feasibility study, which defines the problem, collecting data and analyzing it so
   solutions can be recommended
   preliminary network design, which begins to locate the parameters and provides
   those interested with the opportunity to participate
   detailed network design, which generates specifics on costs and all the other
   areas suggested by the above questions
   implementation, which includes issuing requests-for-proposal, site preparation,
   and training

8. **What are the estimates for overloading and obsolescence?**
   project network traffic forward, estimating where and when overload problems
   will probably occur
   obsolescence cannot be avoided, so assure timely recovery of the network investment (e.g., within so many months)

This data system is for a hypothetical company with many customer accounts, such as a retailer or bank. To learn her account status, a customer telephones the operator, who keys in the account number on a terminal and reads the response that appears on the display screen. Only one inquiry at a time can be processed.

Figure 11–5 illustrates the elements that contribute to the response time:

- the customer calls the operator and asks the status of her account
- the operator keys in the account number
- the inquiry is transmitted to the central computer

- the computer processes this input
- the disk finds and reads the customer's record
- the computer processes this output
- the record is transmitted to the terminal
- the operator reads the account status to the customer.

Response time and utilization are related. Generally, systems are designed so that the average utilization is 60 to 70% of the peak throughput capacity. At 80% of capacity, response time actually *increases*, since the system responds much more slowly (see Figure 11–6). In other words, delays increase as the load increases. Customer calls will pile (or receive a busy signal), and terminals sharing the communications line to the cen-

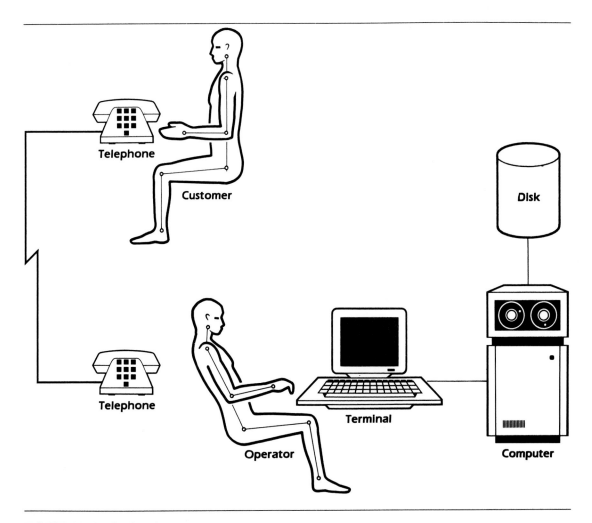

FIGURE 11-4   *On-line data system.*

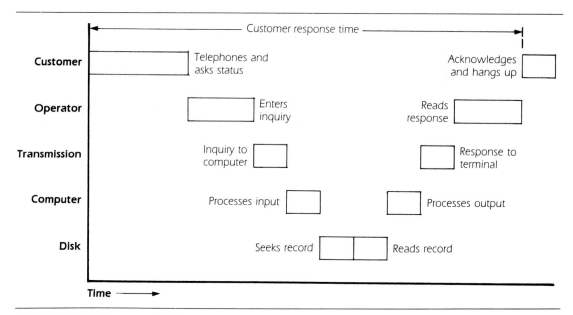

FIGURE 11–5   Response time illustration.

FIGURE 11–6
Utilization versus
response time.

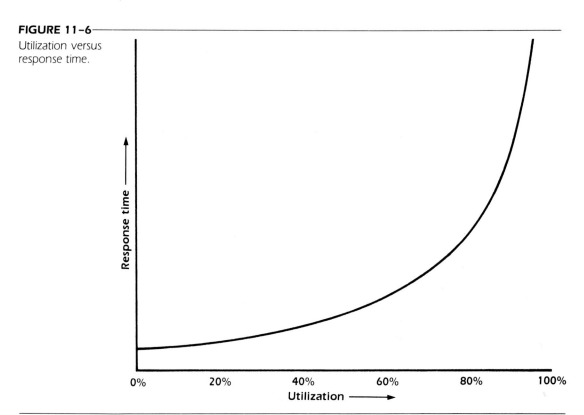

tral computer will have to queue up to use the computer. Bottlenecks can also occur inside the computer (as inquiries wait to be processed) or disk storage device.

In designing our system, we will want to know how well it performs, especially its response times for various loads. Our analysis will show us how the delays build up and where we can expect to find bottlenecks. The system will contain the following components (see also Figure 11–4):

- terminal
- communication line
- central computer
- disk storage device

Let's start estimating some times for the steps in Figure 11–5:

- *Operator enters account number.* After the operator answers the phone, we will say twenty seconds are needed to key in the customer's account number.
- *Transmission to computer.* Assume the customer's account number has 12 digits which will be sent asynchronously at 1200 bits per second in EBCDIC. Each eight-bit EBCDIC character has one start bit and one stop bit, so 10 bits are required to transmit one character. Therefore, the buffered terminal can send 120 characters per second over the communication line to the computer:

$$\frac{1200 \text{ bps}}{10 \text{ bits per character}} = 120 \text{ characters per second}$$

Since the account number has 12 characters, it will require 0.1 second to transmit.

- *Computer processing.* An application program inside the computer calculates the file location from the customer account number. Assume this takes 0.002 second.
- *Disk retrieval.* The disk storage device's read/write arm seeks the location of the customer's record and transfers it to the computer's memory, all in 0.05 second.
- *Computer processing.* The computer processes the file data and prepares a reply, again in 0.002 second.
- *Transmission to terminal.* The computer transmits 200 characters to the terminal at 120 characters per second, which takes 1.7 seconds.
- *Operator responds.* The operator reads the account status to the customer, which takes, say, 10 seconds.

So the total customer response time will be:

operator enters account number	20.0	seconds
transmission to computer	0.1	
computer processing	0.002	
disk retrieval	0.05	
computer processing	0.002	
transmission to terminal	1.7	
operator responds	10.0	
total response time	31.854	seconds

We can now estimate if and when each component of our system will get overloaded. First, let's develop some mathematical expressions for the utilization of each component:

Terminal

$$\text{utilization} = \frac{\text{time occupied}}{\text{time available}}$$

$$= \frac{\text{transactions per hour} \times 31.854 \text{ seconds}}{3600 \text{ seconds per hour}}$$

Communication line

$$\text{utilization} = \frac{\text{time occupied}}{\text{time available}}$$

$$= \frac{\text{transactions per hour} \times (0.1 \text{ seconds} + 1.7 \text{ seconds})}{3600 \text{ seconds per hour}}$$

$$= \frac{\text{transactions per hour} \times 1.8 \text{ seconds}}{3600 \text{ seconds per hour}}$$

Central computer

$$\text{utilization} = \frac{\text{time occupied}}{\text{time available}}$$

$$= \frac{\text{transactions per hour} \times (0.002 \text{ seconds} + 0.002 \text{ seconds})}{3600 \text{ seconds per hour}}$$

$$= \frac{\text{transactions per hour} \times 0.004 \text{ seconds}}{3600 \text{ seconds per hour}}$$

Disk storage device

$$\text{utilization} = \frac{\text{time occupied}}{\text{time available}}$$

$$= \frac{\text{transactions per hour} \times 0.05 \text{ seconds}}{3600 \text{ seconds per hour}}$$

Second, we can calculate the maximum number of transactions per hour each component can handle (i.e., 100% utilization):

Terminal

$$\frac{3600}{31.854} = 113.02 \text{ transactions per hour}$$

Communication line

$$\frac{3600}{1.8} = 2000 \text{ transactions per hour}$$

Central computer

$$\frac{3600}{0.004} = 9,000,000 \text{ transactions per hour}$$

Disk storage device

$$\frac{3600}{0.05} = 72,000 \text{ transactions per hour}$$

Obviously, the slowest component is the operator and terminal, followed by the communication line.

Third, since response time usually begins to slow down at 60 to 70% utilization, we will calculate the number of transactions per hour for the operator and terminal and for the communication line at 70% utilization:

Terminal (70% utilization)

$$\frac{3600 \times 0.7}{31.854} = 79.1 \simeq 80 \text{ transactions per hour}$$

Communication line (70% utilization)

$$\frac{3600 \times 0.7}{1.8} = 1400 \text{ transactions per hour}$$

Since the operator begins to get overloaded at about 80 transactions per hour, we should think about adding another operator and terminal as our system gets busier. As the load increases, we should remain alert for any component that is being utilized at 60 to 70% of capacity, since this could slow down the system.

Our hypothetical system is deliberately simplistic, since real systems have many transaction types. Furthermore, the operator cannot handle every call in 20 seconds, since some customer problems are not routine. Finally, some transactions will require more or less time to process.

Real systems require detailed study that is beyond the scope of this chapter. Interested readers are referred to the notes at the end of this chapter and the "Ten Commandments of Network Design" in Figure 11–7.

——————— NETWORK MONITORING —————————————————

Once the on-line data system is installed, it must be monitored to measure performance and solve any problems that crop up (see Figure 11–8). Net-

FIGURE 11–7 ——————————————————————————————————————

Ten commandments of network design.

1. Productivity can be improved by sensible application of information technologies. That is, data bases, data networks, and distributed systems can fundamentally affect the way organizations operate.

2. The key to productivity is to allow users to obtain the information they need for themselves, preferably without programming (i.e., by providing high-level data base languages and dialogue structures for information retrieval).

3. The difference between ad hoc networks and well-designed networks is like night and day.

4. The greatest danger is that incompatible systems will spread (i.e., hardware, software, data), preventing the emergence of an integrated information system.

5. Good design reduces overall system complexity; bad design increases it.

6. Each processing node should be as self-contained as possible and logical connections to other nodes should be minimized.

7. Success hinges heavily on good design of data files and data bases and on co-ordination of the data structures throughout the organization.

8. The sooner an organization builds the data base it needs, the better. If it goes too far with file development, it may not be able to afford the conversion to a data base.

9. Computer networks are enhanced by rigorously defining and adhering to the interfaces between separately developed information systems. When the interfaces are honored, each development group can work autonomously.

10. Distributed, replicated data may create problems (e.g., when updating, during recovery, or in audits), so avoid duplicating the same data.

These are adapted from James Martin, **Design and Strategy for Distributed Data Processing** (Prentice-Hall, 1981).

FIGURE 11–8————————————————————————————

Users of Racal-Milgo CMS (Communications Management Series) monitor and control operating parameters of modems, multiplexers, and network components from a central site.

Photo courtesy of Racal-Milgo.

work analyzers such as Telenex Corp.'s Autoscope are widely used to provide a graphic interpretation of real-time traffic, as well as line performance analysis, error alarms, and diagnostics for trouble-shooting. The terminal displays the on-line status, including response time and line utilization, while the monitoring programs watch for errors and line failures, sounding an audible alarm when they occur. A trace function automatically stores the troublesome session or transaction for review and diagnosis.

Network monitoring and management are discussed in the article "Monitoring Networks Does Not Equal Net Management" at the end of this chapter.

THE BRISFIELD COMPANY

Read the "Private Network" section of the Brisfield Company case (see pages 19–21), and briefly explain the preferred two leased-lines option and comment on its efficiencies.

── **SUMMARY**

1. Networks exist to help the people who use them, so the users' needs are at the heart of network design.
2. Demand and facilities are two sides of the same coin.
3. Demand is based on message types, traffic volumes and peaks, and which locations need information at which times.
4. Facilities include software, transmission channels, and equipment.
5. Coordination is a major consideration for the network designer, and does not occur spontaneously.
6. Performance is measured by response time and throughput.
7. Systems are generally designed for an average utilization of 60 to 70% of capacity.
8. Once 80% of capacity is reached, response time slows down.
9. Response times for various loads should be estimated to help predict system performance and find bottlenecks.
10. Once the system is installed, it must be monitored to measure performance and solve problems.

────────────────────────────────── **REVIEW QUESTIONS**

1. What is the function of demand analysis in network design?
2. Discuss five common message types.
3. Explain the generalized message format in Figure 11–1.
4. What are some of the common carriers that supply communication channels?
5. Discuss Figure 5–1 in terms of the network design concepts covered in this chapter.
6. Explain the function of gateways.
7. Define response time and throughput.
8. How are response time and throughput related?
9. What are the performance trade-offs between response time and throughput?
10. Explain what happens when a system becomes overloaded.
11. How do we estimate response times for various loads?
12. What is monitoring and how is it done?

13. After reading the article at the end of this chapter, explain why monitoring networks does not equal net management.

14. Your boss tells you to study your organization's telecommunications needs and to recommend a network design. How would you proceed?

ENDNOTE

1. Adapted from James Martin, *Future Developments in Telecommunications* (Englewood Cliffs, N.J.: Prentice-Hall, 1977), second edition, 302.

MONITORING NETWORKS DOES NOT EQUAL NET MANAGEMENT

BUDD BARNES

CHERRY HILL, N.J.—Network management, as the term is generally defined, refers to a disassociated collection of monitors, alarms, recorders, and other devices hooked onto the network.

Individually, these act purely as reporters of malfunctions, conditions, and events—it is left to human operators to respond to and manage problems on the network.

Considering the vast geographic reaches, complexities, and operating speeds of modern communications networks, it is an illusion to think that these devices constitute an adequate management system. And given the tremendous importance to business, industry, and government of the torrents of data that move through these networks, that illusion is a dangerous one.

The communications director of a multinational corporation summed up the problem by quipping of his "management system": "It's great. I can know five minutes before anyone else that the network is going to crash. I can see it coming on my monitors."

So there is a big difference between network monitoring and network management. Increasingly, communications managers are rejecting the notion that monitoring a high-speed network is sufficient safeguard against network failure.

Some managers are making use of totally integrated systems—turnkey products that can be structured onto certain network configurations. Others are finding solutions through the installation of various subsystems that are becoming available.

AUTOMATED MANAGEMENT

Briefly described, implementing an effective real-time management and control system means automating procedures that are now conducted at what amounts to the manual level. As in most instances of applying automation, human response capabilities are not sufficient to handle some of today's jobs. This need for "machine assistance" is especially true in the modern high-speed network environment.

A number of features and characteristics distinguish the enhanced, controlled network. One of these is the outward migration of intelligence from the central control site down to the "root" level of the complex.

These automated systems now have distributed intelligence, at least at the nodal level. Many nodes today handle more traffic than total networks used to handle not so long ago.

In more sophisticated networks, there are examples of the extension of intelligence out to the I/O cards. The means of doing this are available as part of the newer network management packages.

The functions that can now be served as automatic features include all of the disaster-avoidance procedures that human operators used to handle in slower days. Such procedures as setting up alternate routes, making topology changes, measuring buffer utilization, and tracking error rates, for example, can now be handled by the management system.

In addition, functions such as record-keeping, accounting, and other administrative activities are inherent system features.

In the modern controlled network, the same type of information is generated as in the networks that rely solely on monitoring. But the big difference is that the network under automated control is able to respond and act on that information with or without operator intervention. The managed network control system, of course, provides alarms to the operators and presents color graphics displays of network conditions. In addition, it presents specific information on demand about any situation under observation. Under these management conditions, there is no need to scan quantities of extraneous data before keying in problem-avoidance instructions.

In the network operating under true management control, operator stations are not limited to a single site. Control can be had from several strategic locations, as the network manager chooses.

Management Information Systems Week, September 18, 1985. Reprinted with permission of *MIS Week*.

The key to this new generation of advanced management technology was summed up by one network administrator who said, "Plan your escape in advance and you'll never have to use it."

The first step in establishing this escape route is to evaluate and use all of the historic information that the network has amassed during its days as a monitored system. Such data as buffer utilization, bit-error rate statistics, terminal activity levels, line usage, and other related data provide a foundation for programming the management control system.

In addition, input information will be required on channel capacities, network topologies, priority settings, and all other data. A full inventory of all network elements and locations is also essential.

Performing these tasks and establishing the means of managing through the use of readily available technology, combined with readily available information, may be the most imperative function in network operations of the future. The time has come for simple monitoring to give way to effective management.

12 TRANSMISSION ERRORS AND SECURITY

MONITORING

CONDITIONING AND EQUALIZING

ERROR HANDLING

ARCHITECTURE AND ERRORS

SECURITY

PRIVACY

AFTER READING THIS CHAPTER, YOU WILL UNDERSTAND THAT:

- A comprehensible telephone conversation can take place on a noisy line, but data transmission has to be virtually error-free.
- Business telecommunications involves many components, which means there are many possible sources of error.
- The network is often monitored from a central computer room where many types of errors can be highlighted and traced.
- Leased lines may be conditioned to reduce error rates and increase transmission speed.
- Many transmission errors are corrected automatically.
- Computer-related fraud is one important aspect of security.
- There are laws which prohibit the disclosure of certain kinds of information.

When you talk on the telephone, the conversation remains intelligible even when you have a very noisy connection that garbles some words. Noise on the line is just a minor inconvenience. When transmitting data, however, it matters a great deal when the contents of a message, packet, or frame are distorted: the loss of just one bit of financial data, for example, could be disastrous if that bit represents a decimal point between numbers. In other words, data transmission has to be virtually error-free.

Transmission involves many components, as illustrated in Figure 12–1, and each one is a possible source of error:

1. operator error
2. terminal hardware or software error
3. cable transmission error
4. terminal controller hardware or software error

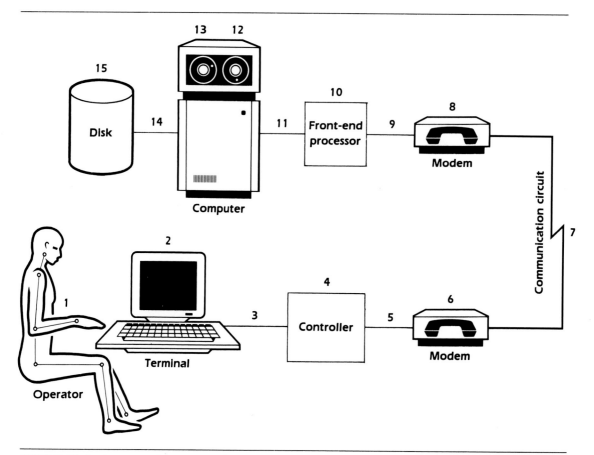

FIGURE 12-1 *Some possible sources of transmission error.*

5. controller port error

6. modem error (at terminal end)

7. communication circuit transmission error

8. modem error (at computer end)

9. front-end processor port error

10. front-end processor hardware or software error

11. transmission channel software error

12. computer hardware error

13. computer software error (e.g., operating system, application programs, data base management system)

14. transmission of software to and from disk

15. disk error

Any of these components can malfunction, and the diagnosis is not always easy. The software people will say it's the hardware and the hardware people will say it's the telephone line and the phone company technician will say it's a software problem. In general, an intermittent problem is probably caused by the hardware, while a recurring difficulty probably originates in the software.

Three important measurements of network performance are:

■ **Mean time between failures** is a measure of the average time within which a given component fails. The user wants to increase this as much as possible.

■ **Mean time to recover** is the average time it takes for the system (e.g., for redundant components) to take over. This can be reduced by rapid diagnosis and correction.

■ **Mean time to repair** is the average time it takes to fix a failed component. This can be decreased by fast repair procedures.

—————— MONITORING——————————————————————————

Data communication lines are often monitored from a central computer room, where intelligent line monitors such as Autoscopes present graphic interpretations of real-time traffic activity, errors and diagnostics, and line performance analysis (see Figure 12–2). Errors are highlighted on the display screen and trigger audible alarms. The monitors can also trace the history of the session or transaction in which the error occurred, displaying it for the operator. Line performance is monitored continuously and statistics are updated so the data is always immediately available.

Figure 12–3 illustrates an Autoscope analysis of a SDLC/SNA line, and Figure 12–4 shows that an X.25 session could not be established because there was no local confirmation of the incoming call.

The distances involved may contribute to network transmission problems: the greater the delay in error diagnosis, the greater the difficulty in correcting it. Inhospitable weather may also cause errors: above-ground cables and microwave transmissions are both affected by storms and temperature variations.

CONDITIONING AND EQUALIZING

Voice communication channels are subject to various kinds of noise from switching, voltage changes, and other electrical effects. If a burst of noise lasts one millisecond (0.001 second), a line transmitting at 9600 bits per second would lose about ten bits and the message would have to be retransmitted. Attenuation—the decay in signal strength as it travels down the circuit—can also cause problems unless the signal is amplified.

Leasing a line to permanently connect terminals to a central computer is one way to reduce noise, since it bypasses the phone company's switch-

FIGURE 12–2

The Autoscope network line analyzer by Telenex Corp. This sophisticated instrument monitors the communication line, measures performance, and diagnoses errors.

Photo courtesy of Telenex Corp.

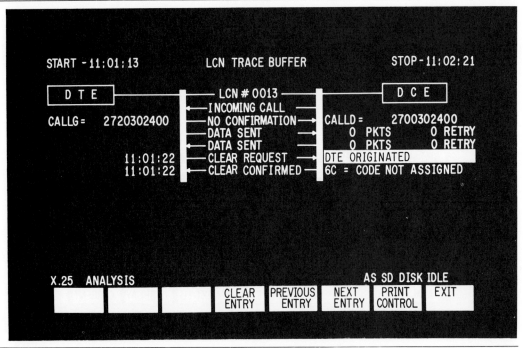

FIGURE 12–3 Analysis of a SDLC/SNA line with an activity report on a 3274-type controller and all terminals connected to it. Detailed reports about response time and other important parameters are indicated on the Autoscope screen.

Photo courtesy of Telenex Corporation.

ing facilities. The leased line can also be *conditioned* to cut error rates and allow higher transmission speeds (up to 9600 bps). This process can only be used with leased lines, since data seldom travels the same path twice over regular phone lines.

Equalization modems allow faster data transmission (up to 4800 bps) over the normal switched telephone network. The modem monitors the line and the incoming signal, and automatically counteracts certain deviations from normal, thereby correcting some circuit-induced distortions.

ERROR HANDLING

When transmission errors occur, they can be detected in various ways— for instance, by parity checks or cyclic redundancy checks. Once the errors are detected, they are corrected automatically and the corrected data is retransmitted by special control characters.

When a terminal sends a block of data over a half-duplex line, it waits for either an ACK or NAK control character from the computer. If an ACK

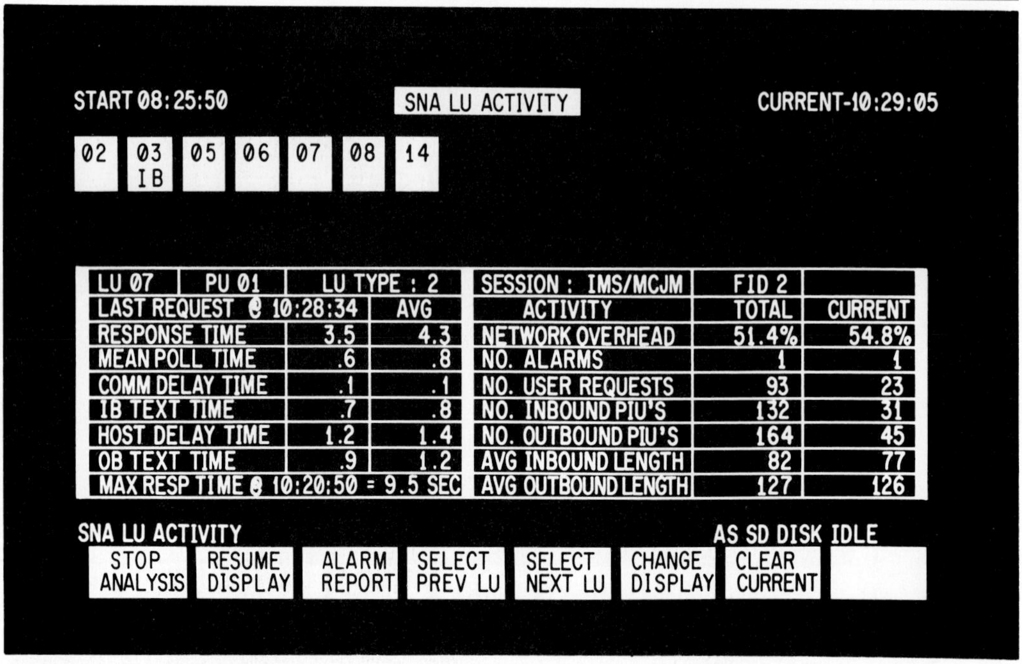

START 08:25:50		SNA LU ACTIVITY		CURRENT-10:29:05	

```
02  03  05  06  07  08  14
    IB
```

LU 07	PU 01	LU TYPE : 2	SESSION : IMS/MCJM	FID 2	
LAST REQUEST @ 10:28:34		AVG	ACTIVITY	TOTAL	CURRENT
RESPONSE TIME	3.5	4.3	NETWORK OVERHEAD	51.4%	54.8%
MEAN POLL TIME	.6	.8	NO. ALARMS	1	1
COMM DELAY TIME	.1	.1	NO. USER REQUESTS	93	23
IB TEXT TIME	.7	.8	NO. INBOUND PIU'S	132	31
HOST DELAY TIME	1.2	1.4	NO. OUTBOUND PIU'S	164	45
OB TEXT TIME	.9	1.2	AVG INBOUND LENGTH	82	77
MAX RESP TIME @ 10:20:50 = 9.5 SEC			AVG OUTBOUND LENGTH	127	126

SNA LU ACTIVITY   AS SD DISK IDLE

STOP ANALYSIS	RESUME DISPLAY	ALARM REPORT	SELECT PREV LU	SELECT NEXT LU	CHANGE DISPLAY	CLEAR CURRENT	

FIGURE 12–4 An Autoscope error diagnosis for an X.25 protocol session that could not be established because there was no local confirmation of the incoming call.

Photo courtesy of Telenex Corporation.

is received (Figure 12–5, left), the terminal sends the next block. If a NAK is received (Figure 12–5, right), the block is retransmitted. Using different acknowledgments for alternating data blocks — ACK-0 and ACKI-1 — insures that blocks cannot be obliterated or lost (see Figure 12–6). For instance, if the second block is lost, the computer repeats ACK-0, which tells the terminal to retransmit the block. The computer than acknowledges with ACK-1, so transmission continues normally.

———— ARCHITECTURE AND ERRORS ————————————

The seven-layer OSI model for network architecture (see Chapter 7) has extensive built-in controls.[1]

Physical control. This layer connects, maintains, and disconnects the physical link between the terminal and the computer. Physical access to the network is restricted at this level.

Data link control. This layer breaks down incoming data into data frames, which are transmitted sequentially. Controls number and count the frames, detect errors, and retransmit the blocks.

Network control. This layer accepts messages from the host computer, converts them into packets, and directs them to their destinations. Controls route the packets, for example, so that one channel is not overloaded with too many packets.

Transport control. This layer accepts whole messages from the session control layer, divides them up, passes the pieces to the network control layer, and makes sure the pieces all arrive at their destination correctly. In the first three layers controls are carried out by each machine, but in this layer the controls are in both the source and destination hardware and software.

Session control. This layer is the user's interface with the network, and it allows the user to log in by typing a password. Controls here protect data base management systems from aborted entries (which would leave inconsistencies in the data base).

Presentation control. This is the layer that makes incompatible terminals work together, modifying line and screen length, character sets, and file formats. Any encryption or security software might be in this layer.

Application control. This layer is defined by the user or the application program, and usually contains controls typical of business information systems (e.g., input data checks).

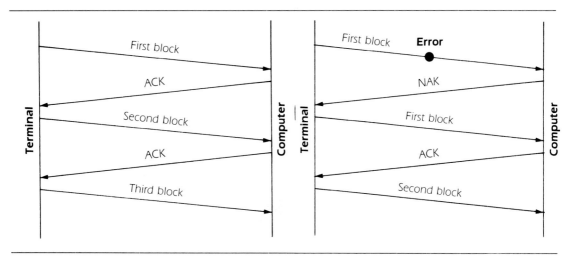

FIGURE 12-5 Half-duplex, block-by-block transmission.

FIGURE 12–6

Alternating
acknowledgments.

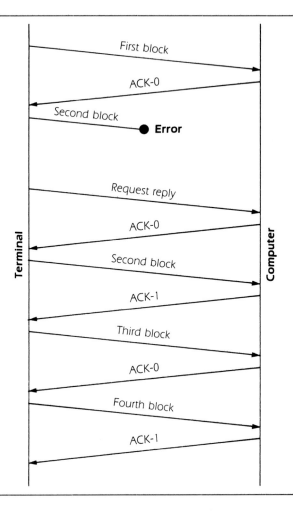

—— **SECURITY** ——————————————————————————————

A hostile employee who knows how can create a software time bomb that
changes records, destroys files, and then self-destructs so the damage remains but its source does not. Since business telecommunications vastly
extends the power of data processing, it also increases the scope for this
kind of villainy (see the article "The Chase Is On for a Telex from Bogota"
at the end of the chapter). Financial institutions are now protecting their
electronic funds transfers with encryption devices (see Figure 12–7), and
the courts are already grappling with the legal issues raised by computer-
related fraud.

In the context of business telecommunications, security has many facets.[2] Physical security, for example, includes fire prevention, insurance coverage, and the locks on the computer room door. Hardware security typically involves a backup power source, file recovery procedures, and preventive maintenance. Software security can mean backing up disks, checking file labels, and preventing user modification of programs.

The following steps help to insure that deliberate or accidental errors are quickly detected and corrected:

- segregation of duties, so that no one person has complete control over a transaction
- system controls, such as input validity and total checks
- control procedures, such as proper supervision of those who verify the work of others.

——— PRIVACY ————————————————————————

The privacy issues inherent in electronic data processing have been a legal concern since passage of the Privacy Act of 1974 and the Freedom of Information Act. Business telecommunications makes it much easier to col-

FIGURE 12-7

The Racal-Milgo Datacryptor 64 data link encryption device offers diagnostic and public key management capabilities, and provides data security for both government and commercial applications.

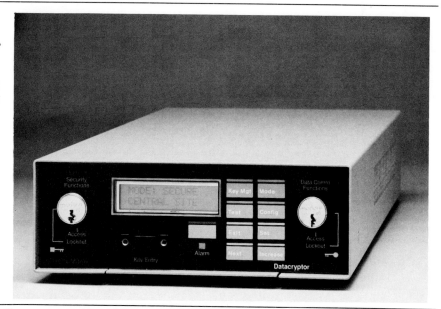

Photo courtesy of Racal-Milgo.

lect, maintain, and disseminate information about an individual's education, medical history, financial transactions, and employment, but such information is subject to the privacy laws.

The Privacy Act of 1974 specifies the instances where:

- records may be disclosed
- individuals must be notified of disclosure
- individuals must be allowed to inspect their records
- individuals may file statements of disagreement to be forwarded with subsequent disclosures.

The Freedom of Information Act, on the other hand, provides for the disclosure of certain kinds of information. Generally, it requires government agencies to promptly make available all reasonably described records that are requested. Exceptions include personnel and medical files which, if disclosed, would "constitute a clearly unwarranted invasion of personal privacy."

Accordingly, business telecommunications users need to know how these legal requirements affect their own telecommunications networks. Access to files, for example, is restricted by the Privacy Act of 1974. The article "Protecting Privacy and Security in the Micro Environment" at the end of this chapter emphasizes the need for sensitivity in this area.

SUMMARY

1. Data transmission has to be virtually error-free, in contrast to telephone conversations.
2. Business telecommunications involves many components: people, hardware, software, and communication lines. Each is a possible source of error.
3. Three measurements of network performance are mean time between failures, mean time to recover, and mean time to repair.
4. Business telecommunications networks are often monitored from a central computer room where line monitors show traffic, errors, and diagnostics.
5. Leased lines can be conditioned to reduce error rates and allow higher transmission speeds.
6. Transmission errors are normally corrected automatically.
7. The seven-layer OSI model has extensive built-in controls.
8. Security insures that deliberate or accidental errors are detected and corrected, and is increasingly important in business telecommunications.
9. The privacy of some kinds of electronic information is protected by law.

REVIEW QUESTIONS

1. Discuss the sources of transmission errors.
2. Explain why error diagnosis may sometimes be difficult.
3. What are three measurements of network performance?
4. What is network monitoring, and how is it done?
5. How are distance and transmission problems related? How about weather?
6. Where is conditioning used?
7. Where are equalization modems used?
8. Explain why ACK-0 and ACK-1 control characters are alternated.
9. The seven-layer OSI model has extensive controls. Comment.
10. Electronic funds transfers are not trouble-free. Comment.
11. How does privacy affect business telecommunications?
12. Personal computers raise questions about information security. Comment.

ENDNOTES

1. For a good discussion, see Jerry FitzGerald, *Business Data Communications* (Wiley, 1984), Chapter 8.
2. See Frank Greenwood and Lee A. Gagnon, *Assessing Computer Center Effectiveness* (New York: American Management Association, 1977), 23–25.

THE CHASE IS ON FOR A TELEX FROM BOGOTA

Every day, international banks receive thousands of instructions by telex and telephone telling them to transfer money all over the place. Do they check carefully enough that the person giving the instructions is who he says he is? The Colombian government thinks not. It is suing Chase Manhattan in a London court for the return of $13.5 million which, it alleges, was fraudulently transferred from its account with the bank in May 1983.

The case is bizarre. In early 1982, Columbia sought around $70 million to buy equipment for its armed forces and police. Chase Manhattan was willing to put together a syndicated loan but, because it has a policy of not lending money for arms purchases, the loan was split into two parts. Just over $21 million was provided for the ministry of defense's budget by a syndicate led by West Germany's Berliner Handels-und-Frankfurter Bank. Another $47 million came from a syndicate led by Chase Manhattan's London branch.

By April 1983, $13.5 million of the Chase loan was still unused. The Colombian finance ministry asked the bank if it would extend the period in which the money could be drawn from one to two years. The ministry was unwilling to pay the extra 1% fee which Chase demanded, so the $13.5 million was drawn on May 10, 1983, and placed in a deposit account with Chase Manhattan in London. The next day, Chase received a telex, apparently from Colombia's central bank and ostensibly in the name of the then director general of public credit, Mr. Jorge Serpa.

The telex instructed Chase to transfer the $13.5 million to an account at Morgan Guaranty in New York. This Chase did. The account at Morgan was a clearing account for the Zurich branch of Israel's Bank Hapoalim. The ultimate owner of the account was Mr. Robert Russell, a Texan who often acts as a middleman for banks in arms and other sensitive deals.

Chase says it immediately informed the Colombian central bank that it had made the transfer of $13.5 million. It heard nothing more about the transfer for five months until, in October 1983, a junior official in the finance ministry, Mr. Diego Dominguez, asked for a statement of the account.

The Colombian government says that the telex to Chase ordering the transfer was fraudulent, and that it received no confirmation from Chase. It also claims that the telex about the $13.5 million was sent directly to Chase in London while all the other (legitimate) drawings on the account went through Chase Manhattan in Bogota. But the Colombians did not use a "test key" with the account—a secret code which banks provide to their customers to safeguard against fraudulent telexes.

In April 1984, Colombia's attorney general listed a dozen suspects he claimed were implicated in fraudulently withdrawing the $13.5 million. Investigations in Colombia have been accompanied by several deaths. Mr. Diego Dominguez was killed in a car crash in December 1983. In November 1984, Mr. José Antonio Vargas, the head of a parliamentary commission investigating the affair, was found riddled with bullets in his flat four days before he was due to report to parliament. Documents detailing his findings had been taken from his study. Shortly afterwards, the lawyer in Bogota appointed by the Colombian government to represent Mr. Russell was shot. The computerized records of telex messages, kept automatically by the telephone company in Bogota, were found to have been conveniently erased for May 11, 1983.

Mr. Russell says that he first became involved in late 1982, when he was approached by a Colombian military contact and asked to act as a middleman in a deal to buy "German arms" for Argentina. A fortnight after the $13.5 million went into Mr. Russell's account, he says he met by arrangement a Colombian calling himself Colonel Lara. Mr. Russell says that acting in good faith he gave this man authority to withdraw $12.7 million from his account at Bank Hapoalim. The balance stayed in the account as Mr. Russell's commission; the $12.7 million went to an account at another Israeli bank, Bank Leumi, in Panama. There the trail ends

The Economist, April 13, 1985, 80. © 1985 *The Economist*, distributed by *Special Features*.

in the mists of Panama's bank secrecy laws.

During the investigations, Colombia's deputy attorney general, Mr. Jaime Hernandez, said that similar frauds had occurred before though the Colombian central bank had not made them public. A parliamentary investigation last year discovered that since 1977 a total of $562 million of foreign loans to Columbia had not been entered into the national budget—more than 10% of all Colombia's new foreign borrowing over the seven years. More misunderstanding over telexes?

PROTECTING PRIVACY AND SECURITY IN THE MICRO ENVIRONMENT

ALAN F. WESTIN

Early this year I visited a major American manufacturing firm whose engineering, sales, and administrative staffs are heavy users of personal computers. A company executive explained, "We're providing our professionals with a powerful analytic tool that they can use at any moment they need it—at the office, at home, or on the road. We want them to be experimental and to have computing power at their fingertips."

However, this firm has become increasingly concerned with threats to the confidentiality and security of company data. Within the past year company officers discovered that employees were taking terminals home, accessing the mainframe with an ID code, and storing their personal files on the mainframe system—all in violation of company rules. They also found that an employee was authorizing complimentary product orders from a microcomputer that bypassed the regular sales order system controls. In addition, a full account of the access procedures for getting into the company telecommunications system suddenly appeared on a hacker's bulletin board.

The experiences and dilemma of this company are typical among business, government, and non-profit organizations. We found the same concerns among 110 organizations examined in a recently completed two-year study: Project on the Workplace Impact of Using VDTs in the Office. All of the surveyed managers reported experiences of and worries about breaches of confidentiality and security.

At several newspapers we visited, management gave reporters a private file on which they could store confidential notes and sources and told them that no one else could get into their file. In fact, the security was so weak that other reporters and editors soon learned how to browse these private lines.

At a law office where attorneys use both a dedicated word processor and a micro system, the office manager became concerned about the security of highly sensitive client data. After creating an empty file and marking it "Confidential," the office manager found that during the first week of its existence, eight attempts had been made to access the file from inside the firm.

At a high-technology company, a manager made adjustments in the electronic mail system so that all messages to other executives but not to him appeared on his terminal. He said he "just wanted to keep informed about what people were saying to one another."

At one bank, personnel officers with personal computers were creating their own automated files on employees. They recorded matters of discipline, lateness, absentee counts, and other matters, especially when they thought "potential Equal Employment Opportunity (EEO) issues might arise." This violated the bank's employee privacy rules and could also have been a serious problem in EEO or unjust discharge cases.

In at least a dozen organizations we visited, company officers told us that professional employees had downloaded proprietary information or valuable customer lists from the mainframes onto floppy disks, which they had taken off the premises and sold to competitors.

Although we found that most management information services executives and security professionals are concerned about these issues, more than 90% of the organizations we visited from 1982 to 1984 had not:

- done a risk assessment of new privacy and security exposures in their organizations
- conducted a sensitivity analysis to classify types of information handled in the micro environment
- issued any policy statements or guidelines to instruct end users in their responsibilities
- trained end users in safeguarding sensitive personal or proprietary data.

Fortunately, this seems to be changing. In 1985 when we sampled some of the organizations visited earlier, we found that about half had created privacy and security task forces to draw up policy guidelines for micro users, issued handbooks on microcomputer confidentiality and security techniques, and set specific responsibilities for managers to assure that their employees will comply with privacy and security policies.

It is clear that inexpensive desktop computers and powerful off-the-shelf software represent a resource that can unlock the creativity of millions of professional and managerial users. But with these powerful tools comes duty of trusteeship for sensitive personal and proprietary information. Any organization that does not address this issue now and does not provide data security training for its microcomputer users is risking serious future trouble.

CONCLUDING NOTE

By now, the reader will probably agree that there is a lot to learn about business telecommunications. This book introduces only the main concepts; obviously, each concept could be (and is) the subject of several books.

As noted in Chapter 9, telecommunications rests on three supports like a three-legged stool: the basic technologies, the regulatory environment, and the products and services available. In this book, we have tried to touch all the bases while emphasizing the relatively unchanging fundamentals of telecommunications. Here are some of the lessons we've learned:

- data processing is an inherent aspect of telecommunications; computers and communications are inseparable
- information networks are moving toward user-controlled applications
- public and private networks are becoming indistinguishable.

One of the important telecommunications issues not discussed in this book is the increasing distance between the public policy makers and those who plan and develop integrated information networks. The technology is evolving so fast that few policy makers understand what is happening. The old regulatory approach that worked for telegraph and telephone companies in the 19th Century now focuses too much attention on the services and too little on the network itself.

Another important issue is related to the breakup of the Bell System. Should the seven regional Bell Operating Companies, which control most of America's local telephone communications, be allowed to compete in equipment manufacturing, long-distance services, and enhanced services such as data bases? If so, then regulated and deregulated businesses would exist side-by-side.

Although many questions have still to be answered, and many details have yet to be worked out, telecommunications will certainly be the key technology of the information age, and it has already made office and factory automation a reality.

GLOSSARY

Access time. The interval from when data is requested until delivery is completed.

Acoustic coupler. A device that converts digital data signals to audible tones for transmission over telephone lines.

Address. A group of digits that tell the computer where certain information is stored.

Alphanumeric. Using both letters and numbers.

Amplitude modulation (AM). A method used to impress information signals on a carrier wave by varying the strength or amplitude of the wave according to the pattern produced by the signals.

Analog. Data in the form of continuously varying physical quantities (*see* Digital).

Application program. A program written for or by a user, or a program used to connect and communicate with stations in a network, enabling users to perform application-oriented activities.

Applications software. Software programs (usually off-the-shelf) which usually allow the user to perform tasks such as word processing.

Archive. A procedure for transferring information from on-line storage media to off-line or external storage (e.g., a diskette or magnetic tape).

ASCII. American National Standard Code for Information Interchange, a standard code consisting of seven-bit coded characters (eight bits including parity check).

Assembler. A device that translates assembly language programs into machine language.

Asynchronous. A method of transmitting data in which each individual character is bracketed by a start and stop bit signal (since each character has at least two additional signal bits, asynchronous transmission is less efficient than synchronous transmission).

Attenuation. The reduction in the strength of an electrical signal that occurs during transmission.

Automatic answer, send, call. A communications feature which allows computers to call, send, and receive text or data without the operator in attendance.

Automation. A production system in which work in process is transferred from one operation to another without human intervention.

Background processing. The automatic execution of a print function (or on some systems, a sort function) while another document is entered or edited.

Background program. In a multiprogram system, a program that can be executed whenever the facilities of the system are not needed by a high-priority program.

Backlog. Incomplete or unprocessed work.

Bandwidth. The difference, expressed in hertz, between the two limiting frequencies of a band.

Baseband. Transmission of a signal at its original frequency, unchanged by modulation.

Baud. The number of times the line condition changes per second; a unit of signalling speed (baud does not necessarily equal bits per second).

Binary. The base two numbering system, which uses only zeros and ones to form all numbers and letters.

Bisync. Binary synchronous communication protocol, a half-duplex communication link protocol.

Bit. A binary digit, the smallest unit of information recognized by a computer (i.e., *zero* or a *one*).

Bits per second (bps). In serial transmission, the speed at which a device transmits a stream of bits.

Black box. Any electronic module that does something to a signal flow.

Blocking/unblocking. To combine records into a block or to select records from a block.

Broadband. A communication channel with a bandwidth greater than a voice channel (i.e., able to handle a greater range of frequencies) and capable of faster data transmission.

BTAM. Basic telecommunications access method.

Buffer. A high-speed storage device that is temporarily reserved for use in input/output operations (often when transferring data from one device to another).

Capacity. Amount of storage space on magnetic media such as cards, cassettes, diskettes, or hard disks (*see* Kilobytes and Megabytes).

Carrier. A continuous frequency signal that is modulated by the patterns produced by a separate information signal.

CCITT. Comité Consultaif Internationale de Telegraphique et Telephonique.

Central processing unit (CPU). The main computer, where data actually gets processed.

Centralization. A system design in which all equipment is centrally located in one area, which allows for more control over the cost and operation of equipment (*see* Decentralization).

Channel. A path used for the transmission of electrical signals; in broadband transmission, one of several available communication links.

Character set. The total number of displayable characters, including alphabetic, numeric, and special symbols.

Chip. A small integrated circuit which may contain many hundreds or thousands of discrete components (transistors, resistors, capacitors, diodes) and anywhere from 4 to 100 connecting pins.

Circuit switching. (See Line switching.)

Cluster controller. A device that controls the input/output operations of multiple devices attached to it (also known as terminal controller).

Coaxial cable. A cable consisting of an outer conductor and an inner conductor, separated by a dielectric.

Code. The representation of information in a form different from its original form that can be understood by both the sender and receiver.

Common carrier. An organization licensed to provide telecommunications facilities to the public.

Communicate. The ability of one computer to talk to another computer or other devices (*see* Compatibility; Teleprocessing).

Communications controller. A unit which manages line control and the routing of data through the network to the host computer.

Communications facilities. Telephone lines, cables, microwaves, satellites, fiber optics, and other devices used to transmit data.

Compatibility. Common characteristics which permit one machine to accept and process data prepared by another machine without conversion or code modification (*see* Communicate).

Concentrator. A device which routes information from multiple communications links into a smaller number of higher-capacity links.

Configuration. The components (controller, display, printer, software, and application programs) which make up a system.

Contention. A method of control that determines how the separate nodes of a network can access a shared transmission medium.

Control character. A coded character which does not print but instead initiates a machine function such as carriage return, centering, underlining, or boldface.

Controller. A logical interface between an input/output device and a computer which performs buffering and code conversion.

Conversational mode. Communication between a terminal and a computer in which each entry from the terminal elicits a response from the computer, and vice-versa.

Cps. Characters per second, measure of printing speed.

CPU. Central processing unit, the main computer.

CRT. Cathode ray tube, a television-like device used to display text and data.

CSMA/CD. Carrier Sense Multiple Access with Collision Detection, an access protocol in which communicating devices compete or contend for a shared communication line.

Cyclic redundancy check (CRC). Error checking performed at both the sending and receiving station after a block check character has been accumulated.

Data base. A collection of related data items or records that is keyed in once and then used by one or more applications.

Data base management system (DBMS). A systematic approach to storing, updating, and retrieving information which allows many users access to the information.

Data communications. The transmission and reception of encoded information over telecommunications lines (see Communicate).

Data entry unit. Generally, a terminal where an operator enters data.

Data link. A connection between one location and another that allows the transmission of data.

Data processing. The execution of a programmed sequence of operations upon stored data; a generic term for business applications.

Decentralization. A systems design in which equipment is scattered in various locations throughout a company (see Centralization).

Digital. Data in the form of digits (see Analog).

Disk, diskette. Types of magnetic storage media (e.g., hard disks, mini-diskettes).

Display. A visual image, usually projected on a cathode ray tube.

Display buffer memory. The buffer holding area for characters displayed on the screen (the size of the buffer may be larger than the number of characters displayed to allow for scrolling).

Distributed data processing. The movement of information processing functions from a central computing facility to separate locations equipped with independent systems.

Distribution. Delivery of output, such as a document or message, to its final destination.

Downtime. Time during which the equipment is not working.

Duplex. A transmission circuit that permits simultaneous two-way communication (also called full-duplex).

EBCDIC. Extended binary-coded decimal interchange code, character set consisting of eight-bit coded characters.

Editing. Correcting, changing, updating, or revising data.

Electronic Industries Association (EIA). A U.S. standards organization specializing in the electrical and functional characteristics of interface equipment.

Electronic mail. The generation, transmission, and display of information by electronic means.

Emulation. The imitation of one system's code set by another so that the two may communicate (for instance, a system with teletype emulation will act like a teletype when communicating with another teletype).

External storage. Devices outside the system which can store information (e.g., disks or magnetic tape).

Facsimile machine. A device that transmits alphanumeric and graphic data to remote sites over telephone lines.

FCC. Federal Communications Commission.

Fiber optics. A medium for data communication that uses light and glass fibers.

Field. A unit of information within a record.

File. A related group of records.

File organization. The manner in which files are arranged, formatted, and accessed on storage media.

Firmware. Software instructions that have been permanently stored in a computer's memory.

Foreground processing. A system dedicated to performing one function at a time (see Background processing).

Frequency division multiplexing (FDM). The division of a communications line into separate frequency bands, each capable of carrying information signals.

Frequency modulation (FM). A method used to impress information signals on a carrier wave by varying the frequency of the wave according to the pattern produced by the signals.

Frequency shift keying (FSK). A method of modulating a carrier signal by shifting its frequency up or down from a mean value (frequency shifts occur when there is a change from one binary value to another).

Front-end processor (FEP). A network processor which provides communications management services to a central computer to which it is attached.

Full-duplex (FDX). (See Duplex.)

Gateway. A communication interface (usually a computer) connecting two or more networks.

Half-duplex (HDX). A transmission circuit that permits alternate two-way communication.

Hardcopy. Machine output, such as a report or listing printed on paper (see Soft copy).

Hard disk. A permanently enclosed rigid platter that offers greater storage capacity and faster access to information than floppy diskettes.

Hardware. The mechanical or electronic equipment used by a system.

Hardwired. Using wired circuitry that cannot be removed (cheaper than software-based systems, but also less flexible).

Hertz (Hz). A unit of measurement of the frequency (number of cycles per second) of an energy waveform.

High-level data link control (HDLC). A standard protocol for data communications defined by the ISO.

Horizontal redundancy check (HRC). Error detection by inserting parity bits in the message and then checking each block for the correct parity (odd or even).

Implementation. Installing equipment and procedures.

Information processing. The entire scope of office automation, including word processing, data processing, image processing, reprographics, and communications.

Information system. A group of computer-based systems and the software required to support one or more business processes.

Input. Data or text to be processed; also, the transfer of data from a keyboard or external device to an internal storage device.

Input device. A terminal or other device which converts data into electronic signals that can be interpreted by computers and related equipment.

I/O. Input /output.

Integration. The combination of data processing, word processing, and telecommunications into a single system.

Intelligent terminal. A terminal with some logic and data-processing capability.

Interactive operation. An on-line operation where there is a real-time give-and-take between a person and a machine (also called conversational mode).

Interexchange circuits (IXC). Telephone company circuits connecting central switching offices.

Interface. An electrical connection used to attach a peripheral device to a system.

Interrupt. In data transmission, to take any action at the receiving station that causes the transmitting station to terminate the transmission.

ISDN. Integrated services digital network.

ISO. International Standards Organization.

Keyboarding. Entering information on a keyboard.

Kilobyte (K). 1024 bytes of information.

LAN. Local area network.

Leased line. A communication channel reserved for a single customer.

LED. Light-emitting diode.

Line switching. A method of completing a physical communication path between two communicating devices (in contrast with message switching, where no physical circuit is established).

Link. A physical medium of transmission, such as a telephone wire, and the associated protocol, communications devices, and programs (also called a line or telecommunications line).

Load. To feed a program into the system.

Local area network (LAN). A communication system that provides economic distribution of information in a local environment.

Macro instructions. A skeletal code that causes the assembler to process a predefined sequence of statements.

Mag card, mag tape. Tape or cards coated with magnetic material on which information is stored.

Mainframe. Another name for the central processing unit, the main computer.

Megabyte (MB). 1000 kilobytes of information.

Menu. A list of functions that an operator can choose from.

Message switching. A communications operation in which messages are received by a switching center and retransmitted to their ultimate destinations.

Microcomputer. A personal computer which uses a microprocessor (usually cheaper and less powerful than either a mainframe or a minicomputer).

Minicomputer. An intermediate range data-processing computer.

Modem. Modulator-demodulator, a device that converts digital signals into analog signals so they can be transmitted over communication lines.

Modulation. The process by which a waveform is varied in accordance with another wave or signal.

Module. An interchangeable plug-in or add-on unit.

Multidrop line. A communication configuration using a single channel or line to serve multiple terminals.

Multifunctional. The ability to perform more than one function.

Multiplexer. A device permitting more than one terminal to transmit data over the same communication channel.

Node. A point in a network where one or more communication lines terminate.

Noise. Unwanted electrical signals that can degrade the performance of a communication channel.

OCR. Optical character recognition, a device which can read printed or typed characters and convert them into digital signals for input into a data- or word-processing system.

Office automation. Using information technologies to do jobs previously performed by office workers.

Off-line. A word- or data-processing operation performed on stand-alone equipment not connected to a central processor or computer system.

On-line. A word- or data-processing operation performed on a local system connected to and sharing the facilities of a remote central processor.

Operating system. Software programs that control the operations of a computer.

OSI. Open system interconnection, ISO's seven-layer model.

Output. The product of an data-processing operation, such as a printed report.

Parity. In ASCII, a final eighth bit that counts the total number of ones in a character.

Password. A word or code that an operator must supply in order to gain access to the system.

Peripherals. Devices such as printers and modems which may be attached to a system to extend its capabilities.

Point-to-point line. A communication line that serves as a direct link between two locations.

Polling. A procedure in which a centralized communications controller contacts several peripheral nodes in turn to allow them time to transmit or receive information.

Port. An access point for data entry or exit.

Printout queuing. A feature which allows documents to be queued for subsequent print while the operator goes on to perform other tasks.

Private branch exchange (PBX). A telephone system serving a specific location that relies on operator intervention to place or receive calls.

Productivity. Output divided by input; a measurement of how well resources are used.

Program. A set of instructions that tell a computer how to perform desired operations.

Protocol. A procedure by which data is transmitted over a communication line, including the format and relative timing of information exchanged between communicating devices.

QTAM. Queued telecommunications access method.

RAM. Random access memory, active memory which is easily changed and erased (information remains in RAM only until the power is turned off).

Recall. To bring information from storage into active memory.

Record. A collection of related data items or fields treated as a unit.

Record processing. The manipulation of information files and the generation of reports from the manipulated data.

Repeater. An amplification device used to restore the original shape and strength of an electrical pulse which has been distorted by attenuation.

Request for proposal (RFP). A commercial solicitation.

Restart. To begin an application program or computer session over again.

RJE. Remote job entry, entering a batch job and receiving output at a remote site.

ROM. Read only memory, static or permanent memory that cannot be changed.

Serial transmission. The sequential transmission of information bits over a single communication channel.

SNA. Systems network architecture, a communications architecture which allows a broad range of terminal types and applications to coexist on the same communication lines and share a network's resources efficiently.

Soft copy. A nonpermanent display of information (e.g., data on a display screen).

Software. The programs and documentation used by a computer system.

Spreadsheet. An electronic worksheet with rows and columns of numbers like a ledger that automatically calculates column and row totals.

Start bit. The first bit of an asynchronously transmitted character that establishes the timing pattern for the receiver.

Stop bit. The last bit or bits of an asynchronously transmitted character that marks the end of that character.

Store-and-forward. The handling of messages in a network by first storing the entire messages and then sending them intact to the next center.

SYN character. The character or characters that precede a block of information in synchronous transmission in order to establish a timing pattern between the sender and receiver.

Synchronous. A method of transmitting data in which a block of characters is bracketed by a single start and stop bit signal.

Telecommunication. (See Teleprocessing).

Telecommunication lines. Telephone and other communication lines that are used to transmit messages from one location to another.

Teleprocessing. Processing data that is received from or sent to remote locations over communication lines.

Terminal. An input/output device such as a cathode ray tube, which can send and receive data.

Terminal controller. A device that controls the input/output operations of multiple devices attached to it.

Time-division multiplexing (TDM). A data communication technique in which communicating devices are assigned specific time slots to access a communication link.

Transponder. A component of a satellite which receives a transmission from earth, amplifies it, changes its frequency, and retransmits it to an earth station.

TSO. Time-sharing option, a method of operation in which a computer facility is shared by several users for different applications at the same time.

Vendor. A company that supplies computer hardware, software, or supplies.

Vertical redundancy check (VRC). Error detection by inserting parity bits in the message and then checking to determine if each character has the correct parity (odd or even).

Virtual circuit. A technique for sending messages in which a logical connection is established between the sender and receiver even though the physical transmission path may vary.

VLSI. Very large-scale integration.

VTAM. Virtual telecommunications access method.

White-collar. Office workers (managers, professionals, and support staff), as opposed to blue-collar factory workers.

Word processing. Hardware and software that process words and text as opposed to numbers.

Work station. A basic physical unit in a system or network which may include a display, keyboard, and storage medium.

X.21. A line communication standard established by the CCITT for synchronous transmission.

X.24. The CCITT standard for interchange circuits between data terminals and circuit-terminating equipment on public data networks.

X.25. The three-layer communication protocol defined by the CCITT as an interface for terminals using public packet-switching networks.

X.400. The CCITT electronic mail standard.

INDEX